FUNDAMENTAL PRINCIPLES OF OCCUPATIONAL HEALTH AND SAFETY

FUNDAMENTAL PRINCIPLES OF OCCUPATIONAL HEALTH AND SAFETY

Benjamin O. ALLI

INTERNATIONAL LABOUR OFFICE • GENEVA

Alli, Benjamin O.
Fundamental principles of occupational health and safety
Geneva, International Labour Office, 2001

Guide: occupational health, occupational safety, ILO Convention, ILO Recommendation, comment, text.
13.04.2
ISBN 92-2-110869-4

ILO Cataloguing in Publication Data

ILO publications can be obtained through major booksellers or ILO local offices in many countries, or direct from ILO Publications, International Labour Office, CH-1211 Geneva 22, Switzerland. Catalogues or lists of new publications are available free of charge from the above address.

Printed in Switzerland PCL

Contents

Annexes

List of boxes

Acknowledgements

My acknowledgements go to all those who contributed to the development and successful completion of this book. Assefa Bequele, Director of the Working Conditions and Environment Department, inspired the author to take on this project. Jukka Takala, Chief of the former Occupational Safety and Health Branch, and Georges Coppée, Head of the former Medical Section, and other colleagues deserve special thanks for reading and making suggestions regarding various drafts of this book. The fact that some are not named individually does not detract from my respect and thanks towards each of them. Angela Haden, Ksenija Radojevic Bovet, John Rodwan and Stirling Smith provided editorial assistance and guidance.

A number of statements, policies or checklists are given in this book. These are intended as illustrations and examples. It is worth emphasizing that the normal ILO disclaimer applies to these texts; they are not authoritative or approved by the ILO.

Introduction

Millions of workers die or are injured or fall ill every year as a result of workplace hazards. The suffering in terms of human life is enormous, while the economic costs of the failure to ensure occupational health and safety are so great that they may undermine national aspirations for sustainable economic and social development.

Improving occupational health and safety is in the best interests of all governments, employers and workers, and measures to make improvements should be discussed and agreed among the ILO constituents. Successful health and safety practice is based on collaboration and good will, as well as taking on board the opinions of the people concerned. It is hoped that this book will provide a basis for action to reduce the high numbers of deaths, accidents and diseases related to working life, and thereby contribute to safe and decent work for all.

As part of its efforts to promote occupational health and safety in its member States, the ILO has compiled comprehensive information on the subject. Much of this information is presented in detail in the ILO's *Encyclopaedia of occupational health and safety.*[1] The principles of occupational health and safety are set out in various Conventions and Recommendations adopted by the International Labour Conference, in particular the Occupational Safety and Health Convention (No. 155), and Recommendation (No. 164), of 1981, and the Occupational Health Services Convention (No. 161), and Recommendation (No. 171), of 1985. Work done by the ILO under the resolution on

[1] Fourth edition, Geneva, 1998; 4 vols; available in print version and as a CD-ROM.

the improvement of working conditions and environment concerning the evaluation of the International Programme for the Improvement of Working Conditions and Environment (PIACT)[2] has also contributed to the improvement of working conditions and the working environment. Today, the ILO's new SafeWork Programme on Safety, Health and the Environment is dealing with security and productivity through health and safety at work, one of its tasks being to produce global statistics on occupational fatalities and injuries.

Based on the abovementioned sources, this book attempts to draw together the fundamental principles of occupational health and safety in a form that will be useful for those involved in policy-making (governments, and employers' and workers' organizations) as well as those within enterprises (managers, supervisors, workers' representatives) who are concerned with the practical implementation of measures to promote and protect the health and safety of workers.

This book aims to serve as a guide or reference for the development of occupational health and safety policies and programmes. It covers the fundamental principles of occupational health and safety, based on the ILO's philosophy of prevention and protection, which stems from the Organization's mandate in the field of occupational health and safety (see box 1). The contents are derived from information embodied in ILO standards, codes of practice and conference resolutions, which outline an occupational health and safety strategy that can be used as a basis for policies and programmes to make the working environment safe and healthy for all. The book summarizes the principles of prevention and protection reflected in these instruments for the use of those responsible for occupational health and safety policies, laws, regulations and programmes.

This publication is primarily intended for policy-makers, workers' organizations, employers and senior labour administrators involved in occupational health and safety activities, who play a pivotal role in ensuring that implementable occupational health and safety policies

[2] Provisional Record No. 37, International Labour Conference, 70th Session, Geneva, 1984 (Seventh Item on the Agenda: Evaluation of the International Programme for the Improvement of Working Conditions and Environment (hereinafter referred to as PIACT)).

are put in place. It is also for the use of factory inspectors, managers, supervisors and workers' representatives in workplaces throughout the world, who implement occupational health and safety policies.

Box 1. The ILO's mandate on occupational health and safety

The ILO's mandate for work in the field of occupational health and safety dates from its very foundation:

> And whereas conditions of labour exist involving such injustice, hardship and privation to large numbers of people as to produce unrest so great that the peace and harmony of the world are imperilled; and an improvement of those conditions is urgently required; as, for example, by the regulation of the hours of work, including the establishment of a maximum working day and week ... the protection of the worker against sickness, disease and injury arising out of his employment....

(Preamble to the *Constitution of the International Labour Organisation*)

This was renewed in 1944, when the relevance of the ILO was reasserted at its Philadelphia Conference:

> The Conference recognises the solemn obligation of the International Labour Organisation to further among the nations of the world programmes which will achieve: ... (g) adequate protection for the life and health of workers in all occupations;...

(*Declaration of Philadelphia,* 1944, para. III)

A single work cannot hope to cover all the subject areas in the vast field of occupational health and safety. This book therefore focuses on the key topics essential to the promotion of occupational health and safety activities. Part I gives an overview of the key concepts which permeate all occupational health and safety activities; Part II presents policy perspectives; and Part III deals with the operational aspects of implementing occupational health and safety.

In 1999, in his first report to the International Labour Conference as Director-General, Juan Somavia declared that "the primary goal of the ILO today is to promote opportunities for women and men to obtain decent and productive work, in conditions of freedom, equity, security and human dignity". This is the concept of decent work.[3]

[3] ILO: *Decent work*, Report of the Director-General, International Labour Conference, 87th Session, 1999, p. 3.

PART I

OVERVIEW

1. Occupational hazards: The problems and the ILO response

An unacceptable situation

Progress in protecting workers' health has always been and continues to be a priority objective for the ILO. Despite immense efforts made since the 1970s, occupational accidents and diseases are still too frequent. Their costs to society and the enterprise, as well as to the workers affected and their families, continue to be unacceptable. There are more than 250 million work-related accidents every year. Workplace hazards and exposures cause over 160 million workers to fall ill annually, while it has been estimated that more than 1.2 million workers die as a result of occupational accidents and diseases. These social costs can no longer be tolerated as the inevitable price of progress. Reducing the toll of occupational accidents and diseases has obvious implications in terms of the alleviation of human suffering.

The related economic costs place a considerable burden on the competitiveness of enterprises. It is estimated that the annual losses resulting from work-related diseases and injuries, in terms of compensation, lost work-days, interruptions of production, training and retraining, medical expenses, and so on, routinely amount to over 4 per cent of the total gross national product (GNP) of all the countries in the world:

- The annual cost of accidents in the manufacturing sector of the United States alone is more than US$190 billion.
- The direct cost of work accidents and diseases is DM56 billion per year in Germany and NOK40 billion in Norway.

- In Australia, the cost of workplace injuries and work-related ill health has been estimated at between A$15 billion to A$37 billion.
- According to the European Agency for Safety and Health, the loss to GNP caused by workplace accidents and work-related ill health in European Union Member States is in the range of 2.6 to 3.8 per cent.[1]

Box 2 illustrates the costs of occupational health and safety in an industrialized country.

Box 2. Costs of health and safety in the United Kingdom

The apex health and safety agency in the United Kingdom, the Health and Safety Executive (HSE), has developed a methodology to calculate the costs of workplace accidents and ill health to:

- individuals;
- employers;
- society as a whole.

The study revealed that, for the year 1995/96 (a 12-month period), costs were:

- for individuals – US$8.96 billion, with costs in future years of US$2.24 billion;
- for employers – US$5.6-11.7 billion, representing 4-8 per cent of gross trading profits;
- for society as a whole – in the region of US$15.8-22.5 billion, equal to around 2 per cent of gross domestic product (GDP); and the average annual growth of GDP in the period 1986-96 was 2.6 per cent.

In other words, the cost of accidents and ill health at work wiped out nearly all economic growth.

Source: *The costs to Britain of workplace accidents and work-related ill health in 1995/96* (London, HSE Books, 1999).

These estimates are sufficient to demonstrate beyond any doubt that a significant reduction in the incidence of occupational accidents and diseases over a reasonable period of time will produce valuable economic benefits. Paying attention to occupational safety

[1] Data provided by the ILO's SafeWork Programme.

and health should therefore be given a high priority not only on moral but also on economic grounds. There is a clear business case for action on health and safety at work.

Variations in performance

There are significant variations in occupational health and safety performance between:

- countries;
- economic sectors; and
- sizes of enterprise.

Countries

The incidence of workplace fatalities varies enormously between countries. There appears to be a significant difference between developed and developing countries:

- a factory worker in Pakistan is eight times more likely to be killed at work than a factory worker in France;
- fatalities amongst transport workers in Kenya are ten times those in Denmark;
- construction workers in Guatemala are six times more likely to die at work than their counterparts in Switzerland.[2]

Economic sectors

Occupational health and safety performance varies significantly between economic sectors within countries. Statistical data show that agriculture, forestry, mining and construction take the lead in the incidence of occupational deaths worldwide. The ILO's SafeWork Programme has estimated, for example, that tropical logging accidents cause 300 deaths per 100,000 workers. In other words, three out of every 1,000 workers engaged in tropical logging die annually or, from a lifetime perspective, on average every tenth logger will die of a work-related accident. Similarly, certain occupations and sec-

[2] World Bank: *World Development Report 1995* (Washington, DC, 1995).

tors, such as meat packaging and mining, have high rates of work-related diseases, including fatal occupational diseases.

Sizes of enterprise

Generally, small workplaces have a worse safety record than large ones. It seems that the rate of fatal and serious injuries in small workplaces (defined as less than 50 employees) is twice that in large workplaces (defined as those with more than 200 employees).

Groups at particular risk

Some groups either seem to be particularly at risk or their specific problems are overlooked. For example:

- The special position of women workers needs attention. The gender division of labour has an impact on women's health and safety in the workplace, which goes well beyond reproductive hazards. As one union points out:

  > Health and safety is male dominated. 86% of Health and Safety Inspectors are male. Resources are traditionally invested far more on "male" industries, rather than areas of industry where women work. Safety standards are based on the model of a male worker. Tasks and equipment are designed for male body size and shape. This can lead to discrimination in a number of areas.[3]

- There are many home-based workers, in both developed and developing countries. Some countries regard them as ordinary employees, subject to normal health and safety legislation. In other countries, they are not included in legislation. But countries that ratify the Home Work Convention, 1996 (No. 177), must ensure protection in the field of occupational health and safety equal to that enjoyed by other workers.
- Part-time workers are another group who may suffer from not being covered by health and safety provisions. This is why the Part-Time Work Convention, 1994 (No. 175), stipulates that "measures shall be taken to ensure that part-time workers receive the same protection as that accorded to comparable full-time workers in respect of: ...

  (b) occupational safety and health" (Article 4).

---

[3] GMB: *Working well together: Health and safety for women* (United Kingdom, 1988).

- Another group that should be considered are contract workers, whose accident rate is on average twice that of permanent workers. Many employers seem to believe that by subcontracting certain tasks, they subcontract their safety responsibilities. This is not the case.
- Drivers are particularly at risk. International estimates suggest that between 15 and 20 per cent of fatalities caused by road accidents are suffered by people in the course of their work, but these deaths are treated as road traffic accidents rather than work-related fatalities.

Despite this worrying situation, international awareness of the magnitude of the problem remains surprisingly modest. The inadequate dissemination of knowledge and information hampers action, especially in developing countries. It also limits the capacity to design and implement effective policies and programmes. The fatality, accident and disease figures are alarming but investment decisions continue to be made in disregard of safety, health and environmental considerations. In the scramble for capital, globalization and increasingly stiff competition tend to deflect attention from the long-term economic benefits of a safe and healthy working environment. While the international press reports on major industrial accidents, the many work-related deaths that occur every day go virtually unrecorded. Workers continue to face serious risks. To reduce the human suffering and financial loss associated with these risks, there is a need for increased and sustained action to protect occupational health and safety, and the environment.

Major occupational health and safety instruments

The means of action used by the ILO to promote occupational health and safety include international labour standards, codes of practice, the provision of technical advice and the dissemination of information. Such action aims to increase the capacity of member States to prevent occupational accidents and work-related diseases by improving working conditions.

One of the main functions of the ILO, from its foundation in 1919, has been the development of *international labour standards*. These cover labour and social matters, and take the form of Conventions and Recommendations. Conventions are comparable to multilateral international treaties which are open for ratification by member States and, once ratified, create specific binding obligations. A government that has ratified a Convention is expected to apply its provisions through legislation or other appropriate means, as indicated in the text of the Convention. The government is also required to report regularly on the application of ratified Conventions. The extent of compliance is subject to examination and public comment by the ILO supervisory machinery. Complaints about alleged non-compliance may be made by the governments of other ratifying States or by employers' or workers' organizations. Procedures exist for investigating and acting upon such complaints.

In contrast, Recommendations are intended to offer non-binding guidelines which may orient national policy and practice. They often elaborate upon the provisions of Conventions on the same subject or a subject not yet covered by a Convention. Although no substantive obligations are entailed, member States have certain important procedural obligations in respect of Recommendations, namely, to submit the texts to their legislative bodies, to report on the resulting action, and to report occasionally at the request of the ILO Governing Body on the measures taken or envisaged to give effect to the provisions.

Conventions and Recommendations adopted by the International Labour Conference, taken as a whole, are considered as an International Labour Code which defines minimum standards in the social and labour field.

ILO standards have exerted considerable influence on the laws and regulations of member States in that many texts have been modelled on the relevant provisions of ILO instruments. Drafts of new legislation or amendments are often prepared with ILO standards in mind so as to ensure compliance with ratified Conventions or to permit the ratification of other Conventions. Indeed, governments frequently consult the ILO, both formally and informally, about the compatibility of proposed legislative texts with international labour standards.

The ILO Conventions and Recommendations on occupational health and safety embody principles which define the rights of workers in this field as well as allocating duties and responsibilities to the competent authorities, to employers and to workers. Occupational health and safety standards broadly fall into six groups, according to their scope or purpose (box 3).

Box 3. Scope and purpose of occupational health and safety standards

Conventions and Recommendations on occupational health and safety may serve several purposes, acting as:

- fundamental principles to guide policies for action;
- general protection measures, for example, guarding of machinery, medical examination of young workers, or limiting the weight of loads to be transported by a single worker;
- protection in specific branches of economic activity, such as mining, the building industry, commerce and dock work;
- protection of specific professions (for example, nurses and seafarers) and categories of workers having particular occupational health needs (such as women or young workers);
- protection against specific risks (ionizing radiation, benzene, asbestos); prevention of occupational cancer; control of air pollution, noise and vibration in the working environment; and measures to ensure safety in the use of chemicals, including the prevention of major industrial accidents;
- organizational measures and procedures relating, for example, to labour inspection or compensation for occupational injuries and diseases.

The ILO policy on occupational health and safety is essentially contained in two international labour Conventions and their accompanying Recommendations (see Annex III for the full text of these instruments). These are:

- ILO Occupational Safety and Health Convention (No. 155), and Recommendation (No. 164), 1981, which provide for the adoption of a national occupational safety and health policy, as well as describing the actions to be taken by governments and within enterprises to promote occupational safety and health and improve the working environment; and

- ILO Occupational Health Services Convention (No. 161), and Recommendation (No. 171), 1985, which provide for the establishment of enterprise-level occupational health services designed to contribute towards implementing occupational safety and health policy.

By the end of December 2000, 35 member States had ratified Convention No. 155, and 19 had ratified Convention No. 161. Their practical impact has, however, been much wider than the number of ratifications would appear to imply, because many countries, although unable to proceed to ratification, have implemented the principles embodied in these instruments.

Boxes 4, 5 and 6 list the most important ILO labour instruments relative to the different aspects of occupational health and safety.

Box 4. Major ILO instruments concerning occupational health and safety in general

Roughly half of the 183 Conventions and 192 Recommendations adopted by the International Labour Conference between 1919 and 2000 address, directly or indirectly, issues of occupational safety and health. General provisions are contained in the following:

- Prevention of Industrial Accidents Recommendation, 1929 (No. 31)
- Protection of Workers' Health Recommendation, 1953 (No. 97)
- Occupational Safety and Health Convention (No. 155), and Recommendation (No. 164), 1981
- Occupational Health Services Convention (No. 161), and Recommendation (No. 171), 1985
- Prevention of Major Industrial Accidents Convention (No. 174), and Recommendation (No.181), 1993

In addition, international labour standards covering general conditions of employment, social security, and the employment of women, children and other categories of workers also have a bearing on health, safety and the working environment. Moreover, a series of Conventions and Recommendations specifically covers the health, safety and welfare of seafarers.

Box 5. Examples of ILO instruments concerning specific risks and substances

- Anthrax Prevention Recommendation, 1919 (No. 3)
- Lead Poisoning (Women and Children) Recommendation, 1919 (No. 4)
- White Lead (Painting) Convention, 1921 (No. 13)
- Radiation Protection Convention (No. 115), and Recommendation (No. 114), 1960
- Guarding of Machinery Convention (No. 119), and Recommendation (No. 118), 1963
- Maximum Weight Convention (No. 127), and Recommendation (No. 128), 1967
- Benzene Convention (No. 136), and Recommendation (No. 144), 1971
- Occupational Cancer Convention (No. 139), and Recommendation (No. 147), 1974
- Working Environment (Air Pollution, Noise and Vibration) Convention (No. 148), and Recommendation (No. 156), 1977
- Asbestos Convention (No. 162), and Recommendation (No. 172), 1986
- Chemicals Convention (No. 170), and Recommendation (No. 177), 1990

Box 6. Examples of ILO instruments concerning health and safety in specific branches of economic activity

- Co-operation in Accident Prevention (Building) Recommendation, 1937 (No. 55)
- Vocational Education (Building) Recommendation, 1937 (No. 56)
- Hygiene (Commerce and Offices) Convention (No. 120), and Recommendation (No. 120), 1964
- Occupational Safety and Health (Dock Work) Convention (No. 152), and Recommendation (No. 160), 1979
- Safety and Health in Construction Convention (No. 167), and Recommendation (No. 175), 1988
- Safety and Health in Mines Convention (No. 176), and Recommendation (No. 183), 1995

2. Key principles in occupational health and safety

A number of key principles underpin the field of occupational health and safety. These principles, which are discussed in detail in subsequent chapters, and the provisions of international labour standards are all designed to achieve a vital objective: that work should take place in a safe and healthy environment.

Core occupational health and safety principles

Occupational health and safety is an extensive multidisciplinary field, invariably touching on issues related to, among other things, medicine and other scientific fields, law, technology, economics and concerns specific to various industries. Despite this variety of concerns and interests, certain basic principles can be identified, including the following:

- *All workers have rights.* Workers, as well as employers and governments, must ensure that these rights are protected and foster decent conditions of labour. As the International Labour Conference stated in 1984:

  (a) work should take place in a safe and healthy working environment;

  (b) conditions of work should be consistent with workers' well-being and human dignity;

  (c) work should offer real possibilities for personal achievement, self-fulfilment and service to society.[1]

---

[1] Conclusions concerning Future Action in the Field of Working Conditions and Environment, adopted by the 70th Session of the International Labour Conference on 26 June 1984, section I, paragraph 2.

- *Occupational health and safety policies must be established.* Such policies must be implemented at both the governmental and enterprise levels. They must be effectively communicated to all parties concerned.
- *There is need for consultation with the social partners* (that is, employers and workers) *and other stakeholders.* This should be done during formulation, implementation and review of such policies.
- *Prevention and protection must be the aim of occupational health and safety programmes and policies.* Efforts must be focused on primary prevention at the workplace level. Workplaces and working environments should be planned and designed to be safe and healthy.
- *Information is vital for the development and implementation of effective programmes and policies.* The collection and dissemination of accurate information on hazards and hazardous materials, surveillance of workplaces, monitoring of compliance with policies and good practices, and other related activities are central to the establishment and enforcement of effective policies.
- *Health promotion is a central element of occupational health practice.* Efforts must be made to enhance workers' physical, mental and social well-being.
- *Occupational health services covering all workers should be established.* Ideally, all workers in all categories of economic activity should have access to such services, which aim to protect and promote workers' health and improve working conditions.
- *Compensation, rehabilitation and curative services must be made available to workers who suffer occupational injuries, accidents and work-related diseases.* Action must be taken to minimize the consequences of occupational hazards.
- *Education and training are vital components of safe, healthy working environments.* Workers and employers must be made aware of the importance and the means of establishing safe working procedures. Trainers must be trained in areas of special relevance to different industries, which have specific occupational health and safety concerns.

- *Workers, employers and competent authorities have certain responsibilities, duties and obligations.* For example, workers must follow established safety procedures; employers must provide safe workplaces and ensure access to first aid; and the competent authorities must devise, communicate and periodically review and update occupational health and safety policies.
- *Policies must be enforced.* A system of inspection must be in place to secure compliance with occupational health and safety and other labour legislation.

Clearly, some overlap exists among these general principles. For example, the gathering and dissemination of information on various facets of occupational health and safety affect all the activities. Information is needed for the prevention as well as the treatment of occupational injuries and diseases. It is also needed for the creation of effective policies and to ensure that they are enforced. Education and training demand information.

While these key principles inform occupational health and safety programmes and policies, the above list is by no means exhaustive. More specialized areas also have corresponding principles. Moreover, ethical considerations regarding such matters as individuals' rights to privacy must be taken into consideration when devising policies.

These basic principles are discussed in the following chapters of this book and in other ILO publications, such as the fourth edition of the *Encyclopaedia of occupational health and safety* (see Introduction, above).

Rights and duties

It is increasingly recognized that the protection of life and health at work is a fundamental workers' right (see box 7); in other words, decent work implies safe work. Furthermore, workers have a duty to take care of their own safety, as well as the safety of anyone who might be affected by what they do or fail to do. This implies a right to know and to stop work in the case of imminent danger to health or safety. In order to take care of their own health and safety, workers need to understand occupational risks and dangers. They should therefore be properly informed of hazards and adequately trained to carry out their tasks safely. To make progress in occupational health and safety within enterprises, workers and their representatives have to cooperate with employers, as well as to participate in elaborating and implementing preventive programmes.

Box 7. Health and safety at work – a human right

The right to safety and health at work is enshrined in the United Nations Universal Declaration of Human Rights, 1948, which states:

> Everyone has the right to work, to free choice of employment, to just and favourable conditions of work....
>
> (Article 23)

The United Nations International Covenant on Economic, Social and Cultural Rights, 1976, reaffirms this right in the following terms:

> The States Parties to the present Covenant recognise the right of everyone to the enjoyment of just and favourable conditions of work, which ensure, in particular: ... (b) safe and healthy working conditions...
>
> (Article 7)

The responsibilities of governments, employers and workers should be seen as complementary and mutually reinforcing to promote occupational health and safety to the greatest extent possible within the constraints of national conditions and practice.

Because occupational hazards arise at the workplace, it is the responsibility of employers to ensure that the working environment is safe and healthy. This means that they must prevent, and protect workers from, occupational risks. But employers' responsibility goes further, entailing a knowledge of occupational hazards and a commitment to ensure that management processes *promote health and safety at work*. For example, an awareness of health and safety implications should guide decisions on the choice of technology or work organization.

Training is one of the most important tasks to be carried out by employers. Workers need to know not only how to do their jobs, but also how to protect their lives and health and those of their co-workers while working. Within enterprises, managers and supervisors are responsible for ensuring that workers are adequately trained for the work that they are expected to undertake. Such training should include information on the health and safety aspects of the work, and on ways to prevent or minimize exposure to hazards. On a wider scale, employers' organizations should instigate training and information programmes on the prevention and control of hazards, and protection against risks. Where necessary, employers must be in a position to deal with accidents and emergencies, including providing *first-aid facilities.* Adequate arrangements should also be made for *compensation* of work-related injuries and diseases, as well as for rehabilitation and to facilitate a prompt return to work. In short, the objective of preventive programmes should be to provide a safe and healthy environment that protects and promotes workers' health and their working capacity.

Governments are responsible for drawing up occupational health and safety policies and making sure that they are implemented. Policies will be reflected in *legislation*, and legislation must be enforced. But legislation cannot cover all workplace risks, and it may also be convenient to address occupational health and safety issues by means of *collective agreements* reached between the social partners. Policies are more likely to be supported and implemented if employers and workers, through their respective organizations, have had a hand in drawing them up. This is regardless of whether they are in the form of laws, regulations, codes or collective agreements.

The competent authority should issue and periodically review regulations or codes of practice; instigate research to identify hazards and to find ways of overcoming them; provide information and advice to employers and workers; and take specific measures to avoid catastrophes where potential risks are high.

The occupational health and safety policy should include provisions for the establishment, progressive extension and operation of occupational health services. The competent authority should supervise and advise on the implementation of a *workers' health surveillance system* and its link with programmes of prevention, protection and promotion of workers' health at the enterprise and national levels. The information provided by surveillance will show whether occupational health and safety standards are being implemented, and where more needs to be done to safeguard workers.

A concise statement that encapsulates the main purposes of occupational health is the definition provided by the Joint ILO/WHO Committee (box 8). As the definition indicates, the main focus in occupational health is on three different objectives:

Box 8. Joint ILO/WHO Committee's definition of occupational health

At its first session in 1950, the Joint ILO/WHO Committee on Occupational Health defined the purpose of occupational health. It revised the definition at its 12th session in 1995 to read as follows:

> Occupational health should aim at: the promotion and maintenance of the highest degree of physical, mental and social well-being of workers in all occupations; the prevention amongst workers of departures from health caused by their working conditions; the protection of workers in their employment from risks resulting from factors adverse to health; the placing and maintenance of the worker in an occupational environment adapted to his physiological and psychological capabilities; and, to summarize: the adaptation of work to man and of each man to his job.

- the maintenance and promotion of workers' health and working capacity;
- the improvement of the working environment and work to become conducive to safety and health; and
- the development of work organizations and working cultures in a direction that supports health and safety at work and thereby also promotes a positive social climate and smooth operation and may enhance productivity of the undertakings. The princi-

ple of working culture is intended in this context to mean a reflection of the essential value systems adopted by the undertaking concerned. Such a culture is reflected in practice in the managerial systems, personnel policy, principles for participation, training policies and quality management of the undertaking.

PART II

POLICY DESIGN AND IMPLEMENTATION

3. Government policy on occupational health and safety[1]

General aims and principles

The promotion of occupational health and safety, as part of an overall improvement in working conditions, represents an important strategy, not only to ensure the well-being of workers but also to contribute positively to productivity. Healthy workers are more likely to have higher work motivation, enjoy greater job satisfaction and contribute to better-quality products and services, thereby enhancing the overall quality of life of individuals and society. The health, safety and well-being of working people are thus prerequisites for quality and productivity improvements, and are of the utmost importance for overall socio-economic, equitable and sustainable development.

In order to ensure that satisfactory and durable results are achieved in the field of occupational health and safety, each country should put in place a coherent national policy aimed at preventing accidents, diseases and injury to health which arise out of, are linked with, or occur during the course of work. By striving to minimize the causes of hazards inherent in the working environment, the policy will reduce the costs associated with work-related injury and disease, contribute to the improvement of working conditions and the working environment, and improve productivity. The articulation of such a

---

[1] This chapter is based mainly on the Occupational Safety and Health Convention (No. 155), and Recommendation (No. 164), 1981, and the Safety and Health in Mines Convention, 1995 (No. 176).

policy will reaffirm a government's commitment to the cause of a safe working environment and enable it to comply with its moral and international obligations by promoting effective action through a unified, coherent and purposeful statement of goals and strategies.

A consistent policy at the national level is particularly necessary in the prevention and control of occupational hazards, where satisfactory and lasting results can only be achieved through sustained and painstaking efforts. There must be coherence in terms of policy content, as well as during the implementation stage. As regards the latter, coherence implies that:

- policy decisions are implemented as rapidly as possible;
- there is continuity of action;
- the measures taken by the various authorities are coordinated and harmonized; and
- resources are made available to enable the necessary measures to be taken.

Box 9. Key features of a national policy on occupational health and safety

- The formulation of the policy should reflect tripartite participation, i.e. there should be inputs from employers' and workers' organizations as well as from government.
- The policy should be consistent with national development objectives and policies as a whole.
- Ways of promoting adequate public awareness and eliciting political support should be envisaged in the policy.
- The policy should include a plan for mobilizing the necessary institutional and financial resources.
- Coordination among all concerned institutions should be fostered as an inherent element of the policy.
- All available means of action should be used consistently.
- The policy should encourage voluntary compliance at enterprise level.
- The policy should be reviewed regularly.

Policy formulation and review

In order to ensure that a national occupational health and safety policy is all-inclusive and comprehensive, measures should be taken to ensure tripartite participation in its formulation, practical implementation and review. Although the substance and approach of these policies can vary according to national conditions and practice, there are nevertheless some basic features that are generally desirable (box 9). An example of tripartite collaboration in formulating a national policy is given in box 10. It should also be borne in mind that if a policy is to be successfully implemented, local conditions and practices must be taken into account when the policy is being formulated.

Improving occupational safety and health is a dynamic process and the objectives are long term. The implementation of any well-thought-out programme may thus be expected to extend over several years. Significant developments or phenomena need to be identified, and the necessary action taken by government as well as within enterprises to avoid a possible disaster. Because the occupational health and safety situation evolves, the policy itself should be reviewed at appropriate intervals. This review may be an overall assessment of the policy or else focus on particular areas.

The objectives of a policy review are to:

- identify major problems;
- devise effective methods of dealing with them;
- formulate and establish priorities for action; and
- evaluate the results.

The nature and the extent of occupational health and safety problems vary from country to country, resulting in part from differences in the level of economic development, and technological and social conditions. For example, while a developing country may still be grappling with the basic occupational health and safety hazards related to agriculture, an industrialized country is possibly confronted with hazards resulting from a modern or advanced technology, such as the health effects associated with the use of visual display units. Similarly, within countries, the incidence of work-related accidents and diseases, including fatal ones, is higher in certain occupations and sectors than in others. Consequently, national policies should establish priorities for action with regard to the specific problems faced within the country concerned. Such priorities might also vary according to the severity or extent of the particular problems, the available means of action, the economic situation of the country, sector or enterprise in question, the effects of changing technology, social conditions, and other factors. It should, however, be stressed that adverse socio-economic conditions must not be used as a pretext for inaction.

Box 10. Tripartite formulation of a national policy on occupational health and safety

The Italian Ministry of Labour unveiled its proposed *Safety at Work Charter* in December 1999 at a national conference. The Charter's contents and aims were worked out jointly with official occupational safety and health agencies, trade unions and employers' organizations.

The Charter sets out to promote the practical application of legislation through three-cornered consultations to identify the best and most efficient ways of preventing work-related accidents and diseases with the highest safety standards for workers.

The government and social partners settled on a joint approach at the conference. The Charter's measures cut across a range of areas:

1. Completing existing legislation and bringing it into line with European Community Directives.
2. Completion of the national health plan 1998-2000 under which a package of health and safety at work information, training, assistance and monitoring measures will be rolled out. Tighter coordination is planned between all relevant government agencies.
3. Incentives for business such as cuts in compulsory employment accident and occupational disease insurance premiums, and training measures for young workers.
4. More workers' safety representatives with a wider role in all workplaces. Smaller firms will be covered by district area workers' representatives.
5. Enforcement machinery will take a more preventive approach, with better circulation of statistics.

Source: *Newsletter of the Trade Union Technical Bureau (TUTB).*

Policy instruments

Given the complexity and the extent of occupational health and safety problems, and the many causes of occupational hazards and work-related diseases, no single intervention would be sufficient in itself to constitute an effective occupational health programme. In order to have an impact, action has to proceed at various levels. The practical measures may vary, depending on the degree of technological, economic and social development of the country concerned, and the type and extent of the resources available. It is possible, however, to give a broad outline of the essential components of a national policy.

In general, a national occupational health and safety policy should provide detailed strategies in the following areas, which will be discussed below:

- national laws, labour codes and regulations;
- role and obligations of the competent authority;
- policy coordination; and
- education and training.

National laws, labour codes and regulations

Appropriate legislation and regulations, together with adequate means of enforcement, are key policy instruments for the protection of workers. They form a basis for efforts to contribute to the improvement in working conditions and the working environment. Occupational health and safety policy should be supported by laws and regulations, along with an inspection mechanism to ensure that these are enforced. The inspection mechanism should make use, amongst other things, of a workers' health surveillance system. The latter could be an integral part of a programme of prevention, protection and promotion run by the government, the community or the enterprise.

Labour legislation lays down minimum standards which are compulsory and applicable to everyone. The enforcement of legislation is fundamental to the implementation of a credible occupational health and safety policy. As employers and plant managers have to fulfil these objectives by adopting appropriate techniques, and as the efficacy of safety measures ultimately rests on their application by workers, it is imperative for representative organizations of employers and workers to be consulted at the various stages in the preparation of laws and regulations.

It has been recognized, in countries with good safety records, that it is more effective to stipulate the duties of those with prime responsibility for occupational health and safety measures in general terms, rather than to attempt to regulate a multitude of hazards in minute detail. This approach is important because technology is developing at an increasingly rapid pace, and it often proves difficult for the legislation to keep abreast of progress. The more recent enactments have therefore avoided setting out detailed requirements, but rather have defined general objectives in broad terms.

The trend in major industrialized countries is to restrict the amount of statutory instruments and to promote the publication by government agencies or specialized professional bodies of directives, codes of practice and voluntary standards, which are more flexible and can be updated more easily. This approach fosters prevention but does not, however, in any way preclude the enactment of specific regulations where strict measures are required to control serious occupational hazards.

Standards, specifications and codes of practice issued by national standards organizations or professional or specialized institutions are generally not binding, but in some cases they have been given the force of law by the competent authority. This practice, which is more common in countries where such organizations and institutions are public, considerably lightens the legislator's task, but it may increase the burden on the occupational health and safety administrations unless they rely on approved bodies or institutions for the application of these standards and specifications.

Role and obligations of the competent authority

Each social partner has a role to play in the formulation and implementation of occupational health and safety policy. The formulation of such a policy should reflect the respective functions and responsibilities of public authorities, employers, workers and others, and it should furthermore recognize the complementary character of those responsibilities. The national authority (designated competent authority) must identify the major problems and draw up a realistic policy, taking into account the resources and means available. It is up to the competent authority to set priorities on the basis of the urgency and importance of the problems to be overcome.

In order to give effect to occupational health and safety policy, and taking account of the available technical means of action, the competent authority or authorities in each country should play certain roles:

- review from time to time the legislation concerning occupational health and safety and any other related provisions issued or approved, e.g. regulations or codes of practice, in the light of experience and advances in science and technology;
- issue or approve regulations, codes of practice or other suitable provisions on occupational health and safety while taking account of the links existing between health and safety on the one hand, and hours of work and rest breaks, on the other;
- undertake or promote studies and research to identify hazards and find means of overcoming them;
- provide specific measures to prevent catastrophes, coordinate and make coherent actions to be taken at different levels particularly where potential high-risk areas for workers and the population at large are situated;
- provide information and advice, in an appropriate manner, to employers and workers and promote or facilitate cooperation between them and their organizations, with a view to eliminating hazards or reducing them as far as practicable;
- ensure that national laws and regulations, as well as other approved provisions are clear, consistent and comprehensive, and reflect national conditions; and
- verify that national legislation takes into account the applicable provisions of international labour standards, especially Conventions Nos. 155 and 161 and their accompanying Recommendations.

With regard to ensuring that the policy is implemented within enterprises, the competent authority or authorities should:

- set the conditions governing the design, construction and layout of undertakings with a view to avoiding or minimizing hazards;
- ensure that hazards are avoided or controlled during the commencement of operations or when major alterations or changes are made;
- verify the safety of technical equipment used at work;
- see to it that the procedures defined by the competent authority are enforced;
- identify work processes, substances and agents which are to be prohibited, limited or made subject to authorization or control, taking into consideration the possibility of simultaneous exposure to several substances or agents;
- establish and apply procedures for the notification of occupational accidents and diseases by employers and, when appropriate, insurance institutions and others directly concerned, and produce annual statistics on occupational accidents and diseases;
- hold inquiries in cases of occupational accidents, occupational diseases or any other injuries to health which arise in the course of or in connection with work and which appear to reflect situations that are serious;
- publish information on measures taken in pursuance of the national policy on occupational health and safety, and on occupational accidents, occupational diseases and other injuries to health which arise in the course of or in connection with work; and
- introduce or extend systems to examine chemical, physical and biological agents as well as ergonomics and psycho-social factors in respect of the risk to the health of workers, bearing in mind that the extent to which this can be done will, of course, depend on national conditions and possibilities.

Policy coordination

In order to ensure coherence in formulating and applying the national occupational health and safety policy, there is a need for coordination between the various authorities and bodies chosen or designated to implement the envisaged actions. There should also be close cooperation between public authorities and representative employers' and workers' organizations, as well as other bodies concerned, in the formulation and application of the national policy. This should include consultation between the most representative organizations of employers and workers, and other relevant bodies, with a view to making arrangements that are appropriate to national conditions and practice. Such arrangements might include the establishment of a central body to take an overall responsibility for implementation of policy measures.

The main purposes of the arrangements referred to above should be to:

- fulfil the requirements regarding policy formulation, implementation and periodic review;
- coordinate efforts to carry out the functions assigned to the competent authority;
- coordinate related activities that are undertaken nationally, regionally or locally by public authorities, employers and their organizations, workers' organizations and representatives, and other persons or bodies concerned; and
- promote exchanges of views, information and experience nationwide, within particular industries, or in specific branches of economic activity.

Education and training

Education and training provide individuals with the basic theoretical and practical knowledge required for the successful exercise of their trade or occupation and their integration into the working environment. Because of the importance of occupational health and safety, measures should be taken to include these subjects in all levels of education and training, including higher technical, medical and professional education. Occupational health and safety training should meet the needs of all workers, and should be promoted in a manner that is appropriate to national conditions and practice.

The idea is to incorporate occupational health and safety principles related to the student's needs into the teaching of all trades and professions. It is therefore important to ensure the integration of occupational health and safety matters, at a level which is in line with the nature of the future functions and responsibilities of the persons concerned, in the curricula and teaching material of all trades and occupations. In general, individuals have great difficulty in modifying acquired habits or abandoning ingrained gestures and reflexes. Schooling or apprenticeship should therefore inculcate safe working methods and behaviour at an early stage, and these will then be followed throughout working life.

Vocational training, whether in the enterprise or at school, often leaves workers poorly prepared to deal with the hazards of their trade. Where they have learnt to work with defective or badly guarded machines and tools, it would be surprising if they were later to be much concerned about safety. If, on leaving school, they are unaware of the importance of good personal hygiene, they are scarcely likely to practise it in the workshop. If people are to be taught how to earn their living, they should also be taught how to protect their lives.

The need to give appropriate training in occupational health and safety to workers and their representatives in the enterprise should thus be stressed as a fundamental element of occupational health and safety policy, and should be stated explicitly in the policy document. Workers should be provided with adequate training in terms of the technical level of their activity and the nature of their responsibilities. Employers should also learn how to gain the confidence of their workers and motivate them; this latter aspect is as important as the technical content of training.

The need to train labour inspectors, occupational health and safety specialists and others directly concerned with the improvement of working conditions and the working environment cannot be overemphasized and should be reflected in the policy document. The training should take into account the increasing complexity of work processes, often brought about by the introduction of new or advanced technology, and the need for more effective methods of analysis to identify and measure hazards, as well as action to protect workers against them.

Employers' and workers' organizations should take positive action to carry out training and information programmes with a view to preventing potential occupational hazards in the working environment, and to controlling and protecting against existing risks such as those due to air pollution, noise and vibration. The public authorities have the responsibility to promote training and to act as a catalyst by providing resources and specialized personnel where necessary. Such support is essential in developing countries.

Occupational health and safety training is a long-term task, and one that is never completely finished. It is a continuous exercise because initial training, even under the best of conditions, cannot cover all foreseeable and unforeseeable situations. If the goals of occupational health and safety policy are to be achieved, there is a need for the continuing involvement of employers and workers. National tripartite seminars can be an effective means of associating employers and workers in the policy-making process. The consensus developed by such seminars increases the commitment to implement the agreed measures.

4. Occupational health and safety policy within the enterprise [1]

General framework

Since occupational accidents and work-related injuries to health occur at the individual workplace, preventive and control measures within the enterprise should be planned and initiated jointly by the employer, managers and workers concerned.

Measures for the prevention and control of occupational hazards in the workplace should be based upon a clear, implementable and well-defined policy at the level of the enterprise. The occupational health and safety policy represents the foundation from which occupational health and safety goals and objectives, performance measures, and other system components are developed. It should be concise, easily understood, approved by the highest level of management, and known by all employees in the organization.

The policy should be in written form and should cover the organizational arrangements to ensure occupational health and safety. In particular, it should:

- allocate the various responsibilities for occupational health and safety within the enterprise;
- bring policy information to the notice of every worker, supervisor and manager;

[1] This chapter is based mainly on the Occupational Safety and Health Convention (No. 155), and Recommendation (No. 164), 1981, and the Safety and Health in Mines Convention, 1995 (No. 176).

- determine how occupational health services are to be organized; and
- specify measures to be taken for the surveillance of the working environment and workers' health.

The policy may be expressed in terms of organizational mission and vision statements, as a document that reflects the enterprise's occupational health and safety values (see box 11). It should allocate the various responsibilities regarding occupational health and safety, including that of bringing policy information to the notice of every worker. The policy should also define the duties and responsibilities of the departmental head or the occupational safety and health team leader who will be the prime mover in the process of translating policy objectives into reality within the enterprise.

The policy document must be printed in a language or medium readily understood by the workers. Where illiteracy levels are high, clear non-verbal forms of communication must be used. The policy statement should be clearly formulated and designed to fit the particular organization for which it is intended. It should be circulated so that every employee has the opportunity to become familiar with it. The policy should also be prominently displayed throughout the workplace to act as a constant reminder to all. In particular, it should be posted in all management offices to remind managers of their obligations in this important aspect of company operations. In addition, appropriate measures should be taken by the competent authority to provide guidance to employers and workers so as to help them comply with their legal obligations. To ensure that the workers accept the health and safety policy objectives, the employer should establish the policy through a process of information exchange and discussion with them. For an example of a health and safety policy, see Annex IV.

Reviewing a policy statement also keeps it alive. A policy may need to be revised in the light of new experience, or because of new hazards or organizational changes. Revision may also be necessary if the nature of the work that is carried out changes, or if new plant or new hazards are introduced into the workplace. It may also be necessary if new regulations, codes of practice or official guidelines relevant to the activities of the enterprise are issued.

Box 11. Checklist for employers writing a safety policy statement

The following checklist is intended as an aid in drawing up and reviewing your safety policy statement. Some of the points listed may be relevant in your case, or there may be additional points that you may wish to cover.

General considerations

- Does the statement express a commitment to health and safety and are your obligations towards your employees made clear?
- Does the statement say which senior officer is responsible for seeing that it is implemented and for keeping it under review, and how this will be done?
- Is it signed and dated by you or a partner or senior director?
- Have the views of managers and supervisors, safety representatives and of the safety committee been taken into account?
- Were the duties set out in the statement discussed with the people concerned in advance, and accepted by them, and do they understand how their performance is to be assessed and what resources they have at their disposal?
- Does the statement make clear that cooperation on the part of all employees is vital to the success of your health and safety policy?
- Does it say how employees are to be involved in health and safety matters, for example by being consulted, by taking part in inspections, and by sitting on a safety committee?
- Does it show clearly how the duties for health and safety are allocated, and are the responsibilities at different levels described?
- Does it say who is responsible for the following matters (including deputies where appropriate)?
  - — reporting investigations and recording accidents
  - — fire precautions, fire drill, evacuation procedures
  - — first aid
  - — safety inspections
  - — the training programme
  - — ensuring that legal requirements are met, for example regular testing of lifts and notifying accidents to the health and safety inspector.

Special considerations

Plant and substances

- Keeping the workplace, including staircases, floors, ways in and out, washrooms, etc., in a safe and clean condition by cleaning, maintenance and repair.
- Maintenance of equipment such as tools, ladders, etc. Are they in safe condition?

- Maintenance and proper use of safety equipment such as helmets, boots, goggles, respirators, etc.
- Maintenance and proper use of plant, machinery and guards.
- Regular testing and maintenance of lifts, hoists, cranes, pressure systems, boilers and other dangerous machinery, emergency repair work, and safe methods of doing it.
- Maintenance of electrical installations and equipment.
- Safe storage, handling and, where applicable, packaging, labelling and transport of dangerous substances.
- Controls on work involving harmful substances such as lead and asbestos.
- The introduction of new plant, equipment or substances into the workplace – by examination, testing and consultation with the workforce.

Other hazards

- Noise problems – wearing of ear protection, and control of noise at source.
- Preventing unnecessary or unauthorized entry into hazardous areas.
- Lifting of heavy or awkward loads.
- Protecting the safety of employees against assault when handling or transporting the employer's money or valuables.
- Special hazards to employees when working on unfamiliar sites, including discussion with site manager where necessary.
- Control of works transport, e.g. fork lift trucks, by restricting use to experienced and authorized operators or operators under instruction (which should deal fully with safety aspects).

Emergencies

- Ensuring that fire exits are marked, unlocked and unobstructed.
- Maintenance and testing of fire-fighting equipment, fire drills and evacuation procedures.
- First aid, including name and location of person responsible for first aid and of deputy, and location of first-aid box.

Communication

- Giving your employees information about the general duties under the HSW Act and specific legal requirements relating to their work.
- Giving employees necessary information about substances, plant, machinery, and equipment with which they come into contact.
- Discussing with contractors, before they come on site, how they can plan to do their job, whether they need equipment of yours to help them, whether they can operate in a segregated area or when part of the plant is shut down and, if not, what hazards they may create for your employees and vice versa.

Training

- Training employees, supervisors and managers to enable them to work safely and to carry out their health and safety responsibilities efficiently.

Supervising

- Supervising employees so far as necessary for their safety – especially young workers, new employees and employees carrying out unfamiliar tasks.

Keeping check

- Regular inspections and checks of the workplace, machinery appliances and working methods.

Source: Health and Safety Executive, United Kingdom: *Writing a safety policy: Guidance for employers* (no date).

Employers' responsibilities

The enterprise policy should reflect the responsibility of employers to provide a safe and healthy working environment. The measures that need to be taken will vary depending on the branch of economic activity and the type of work performed; in general, however, employers should:

- provide and maintain workplaces, machinery and equipment, and use work methods, which are as safe and without risk to health as is reasonably practicable (see box 12);
- ensure that, so far as reasonably practicable, chemical, physical and biological substances and agents under their control are without risk to health when appropriate measures of protection are taken;
- give the necessary instructions and training to managers and staff, taking account of the functions and capacities of different categories of workers;
- provide adequate supervision of work, of work practices and of the application and use of occupational health and safety measures;
- institute organizational arrangements regarding occupational safety and health adapted to the size of the undertaking and nature of its activities;
- provide adequate personal protective clothing and equipment without cost to the worker, when hazards cannot be otherwise prevented or controlled;

- ensure that work organization, particularly with respect to hours of work and rest breaks, does not adversely affect occupational safety and health;
- take all reasonable and practicable measures to eliminate excessive physical and mental fatigue;
- provide, where necessary, for measures to deal with emergencies and accidents, including adequate first-aid arrangements;
- undertake studies and research or otherwise keep abreast of the scientific and technical knowledge necessary to comply with the obligations listed above; and
- cooperate with other employers in improving occupational health and safety.

Box 12. Hierarchy of preventive and protective measures

In taking preventive and protective measures, the employer should assess the risk and deal with it in the following order of priority:

- eliminate the risk;
- control the risk at source;
- minimize the risk by means that include the design of safe work systems;
- in so far as the risk remains, provide for the use of personal protective equipment.

Workers' duties and rights

The cooperation of workers within the enterprise is vital for the prevention of occupational accidents and diseases. The enterprise policy should therefore encourage workers and their representatives to play their essential role in this regard and ensure that they are given adequate information on measures taken by the employer to secure occupational safety and health, appropriate training in occupational safety and health, as well as the opportunity to enquire into and be consulted by the employer on all aspects of occupational safety and health associated with their work.

The enterprise policy should outline the individual duty of workers to cooperate in implementing the occupational health and safety policy within the enterprise. In particular, workers have a duty to:

- take reasonable care for their own safety and that of other persons who may be affected by their acts or omissions;
- comply with instructions given for their own safety and health, and those of others, and with safety and health procedures;
- use safety devices and protective equipment correctly (and not render them inoperative);
- report forthwith to their immediate supervisor any situation which they have reason to believe could present a hazard and which they cannot themselves correct; and
- report any accident or injury to health which arises in the course of or in connection with work.

Workers' duties in hazard control have as their counterpart the recognition of certain basic rights, and these should also be reflected in the enterprise policy. In particular, workers have the right to remove themselves from danger, and to refuse to carry out or continue work which they have reasonable justification to believe presents an imminent and serious threat to their life or health. They should be protected from unforeseen consequences of their actions. In addition, workers should be able to:

- request and obtain, where there is cause for concern on health and safety grounds, inspections and investigations to be conducted by the employer and the competent authority;
- know and be informed of workplace hazards that may affect their health or safety;
- obtain information relevant to their health or safety, held by the employer or the competent authority; and
- collectively select health and safety representatives.

Access to better information is a prime condition for significant, positive contributions by workers and their representatives to occupational hazard control. The enterprise policy should make sure that workers are able to obtain any necessary assistance in this regard from their trade union organizations, which have a legitimate right to be involved in anything that concerns the protection of the life and health of their members.

Health and safety committees

Cooperation between management and workers or their representatives at the workplace, in the field of occupational health and safety, is an essential element in maintaining a healthy working environment. It may also contribute to the establishment and maintenance of a good social climate and to the achievement of wider objectives. Depending on national practice, this cooperation could be facilitated by the appointment of workers' safety delegates, or workers' safety and health committees, or joint safety and health committees, composed equally of workers' and employers' representatives. Workers' organizations play a very important role in reducing the toll of accidents and ill health. One study found that establishments with joint consultative committees, where all employee representatives were appointed by unions, had significantly fewer workplace injuries than those where the management alone determined health and safety arrangements.[2]

The appointment of joint health and safety committees and of workers' safety delegates is now common practice, and can help to promote workers' active involvement in health and safety work. Furthermore, safety delegates are known to be effective in the tasks of monitoring the health and safety aspects of shop-floor operations and in introducing corrective measures where necessary.

Joint health and safety committees provide a valuable framework for discussion and for concerted action to improve safety and health. They should meet regularly and should periodically inspect the workplace. Workers' safety delegates, workers' safety and health committees, and joint safety and health committees (or, as appropriate, other workers' representatives) should be:

- given adequate information on health and safety matters;
- enabled to examine factors affecting health and safety;
- encouraged to propose safety and health measures;
- consulted when major new safety and health measures are envisaged and before they are carried out;

[2] Barry Reilly, Pierella Paci and Peter Holl: "Unions, safety committees and workplace injuries", in *British Journal of Industrial Relations*, Vol. 33, No. 2, June 1995.

- ready to seek the support of workers for health and safety measures;
- consulted in planning alterations of work processes, work content or organization of work which may have safety or health implications for workers;
- given protection from dismissal and other measures prejudicial to them while exercising their functions in the field of occupational health and safety as workers' representatives or as members of health and safety committees;
- able to contribute to the decision-making process within the enterprise regarding matters of health and safety;
- allowed access to all parts of the workplace;
- able to communicate with workers on health and safety matters during working hours at the workplace;
- free to contact labour inspectors;
- able to contribute to negotiations within the enterprise on occupational health and safety matters;
- granted reasonable time during paid working hours to exercise their health and safety functions and to receive training related to these functions; and
- able to have recourse to specialists for advice on particular health and safety problems.

Safety committees or joint health and safety committees have already been set up in larger enterprises in a number of countries.[3] Smaller firms sometimes group together to set up regional health and safety committees for each branch of activity. The most promising results seem to have been achieved when management has concentrated on increasing workers' awareness of their important role in health and safety and encouraged them to assume their responsibilities more fully.

[3] For more details, see ILO: *Encyclopaedia of occupational health and safety*, op. cit.

5. Management of occupational health and safety[1]

The protection of workers from occupational accidents and diseases is primarily a management responsibility, on a par with other managerial tasks such as setting production targets, ensuring the quality of products or providing customer services. Management sets the direction for the company. The strategic vision and mission statement establish a context for growth, profitability and production, as well as placing a value on workers' safety and health throughout the enterprise. The system for managing safety and health should be integrated within the company's business culture and processes.

If management demonstrates in words and action, through policies, procedures and financial incentives, that it is committed to workers' health and safety, then supervisors and workers will respond by ensuring that work is performed safely throughout the enterprise. Occupational safety and health should not be treated as a separate process, but one that is integral to the way in which activities take place in the company. In order to achieve the objective of safe and healthy working conditions and environment, employers should institute organizational arrangements adapted to the size of the enterprise and the nature of its activities.

[1] This chapter is based mainly on the Occupational Safety and Health Convention (No. 155), and Recommendation (No. 164), 1981, and PIACT (op. cit.).

Management commitment and resources

While top management has the ultimate responsibility for the health and safety programme in an enterprise, authority should be delegated to all management levels for ensuring safe operation. Supervisors are obviously the key persons in such a programme because they are in constant contact with the employees. As safety officers, they act in a staff capacity to help administer safety policy, to provide technical information, to help with training and to supply programme material.

Total commitment on the part of management to making health and safety a priority is essential to a successful occupational health and safety programme in the workplace. It is only when management plays a positive role that workers view such programmes as a worthwhile and sustainable exercise. The boardroom has the influence, power and resources to take initiatives and to set the pattern for a safe and healthy working environment.

Management commitment to occupational health and safety may be demonstrated in various ways, such as:

- allocating sufficient resources (financial and human) for the proper functioning of the occupational health and safety programme;
- establishing organizational structures to support managers and employees in their occupational health and safety duties; and
- designating a senior management representative to be responsible for overseeing the proper functioning of occupational health and safety management.

The process of organizing and running an occupational health and safety system requires substantial capital investment. To manage health and safety efficiently, the financial resources must be allocated within business units as part of overall running costs. The local management team must understand the value that corporate leaders place on providing a safe place of work for employees. There should be incentives for managers to ensure that resources are deployed for all aspects of health and safety. The challenge is to institutionalize health and safety within the planning process. Once the programme is under way, concerted efforts must be made to guarantee its sustainability.

Workers' participation

Cooperation between management and workers or their representatives within an enterprise is an essential element of the organizational measures that need to be taken in order to prevent occupational accidents and diseases at the workplace. Participation is a fundamental workers' right, and it is also a duty. Employers have various obligations with regard to providing a safe and healthy workplace (see Chapter 4), and workers should, in the course of performing their work, cooperate in order to enable their employer to fulfil those obligations. Their representatives in the undertaking must also cooperate with the employer in the field of occupational health and safety. Employee participation has been identified as the determinant of successful occupational health and safety management and a major contributing factor in the reduction of occupational diseases and injuries.

The full participation of workers in any occupational health and safety programmes designed for their benefit will not only ensure the efficacy of such measures, but also make it possible to sustain an acceptable level of health and safety at a reasonable cost. At the shop-floor level, workers and their representatives should be enabled to participate in the definition of issues, goals and resulting actions related to occupational health and safety.

Training

The continuous integration of improvements into the work process is vital, but it is possible only if everyone involved is properly trained. Training is an essential element in maintaining a healthy and safe workplace and has been an integral component of occupational health and safety management for many years. Managers, supervisory staff and workers all need to be trained. Workers and their representatives in the undertaking should be given appropriate training in occupational health and safety. It is up to management to give the necessary instructions and training, taking account of the functions and capacities of different categories of workers (see box 13). The primary role of training in occupational health and safety is to promote action. It must therefore stimulate awareness, impart knowledge and help recipients to adapt to their own roles.

Training in occupational health and safety should not be treated in isolation; it should feature as an integral part of job training and be incorporated into daily work procedures on the shop floor. Management must ensure that all those who play a part in the production process are trained in the technical skills that they need to do their work. Training for the acquisition of technical skills should therefore always include a component of occupational health and safety.

Box 13. Management responsibilities in occupational health and safety training

It is the responsibility of management to:

- give each worker practical and appropriate instruction taking account of his or her skills and professional experience, in each case defining the objective to be achieved in terms of ability to perform a specific function;
- provide training involving the acquisition of knowledge and know-how to be applied in a specific job and corresponding to the qualifications required; this may consist of initial training for entry to a particular trade or profession, or adaptive training associated with a modification of the workstation, the introduction of new methods or a transfer to another job;
- give refresher courses to update the knowledge acquired through training; and
- provide further training, thus enabling workers to acquire new knowledge, supplement existing knowledge, or specialize in a particular area by acquiring more detailed knowledge.

Organizational aspects

The control of occupational hazards and diseases requires adequate organizational measures. As there is no perfect model for an organizational structure, a choice has to be made by weighing up the anticipated merits and disadvantages of various systems. Moderation should be the guiding principle, and a step-by-step approach is likely to be more successful than an overambitious scheme that does not allow for subsequent adjustment.

Setting priorities

The first step is to establish priorities among objectives by assessing the main factors contributing to the hazards with the most severe consequences. High priority may also be allocated to actions that will produce rapid results, as early successes will enhance the credibility of efforts. Priorities may change from time to time, depending on the existing situation. It should be reiterated that cooperation between management and workers or their representatives within the enterprise is essential in ensuring the successful implementation of an organizational structure for occupational health and safety.

Planning and development activities

These need to be undertaken both in initially setting up the occupational health and safety management system and in its ongoing revision and modification. Systems and procedures should be thought through logically, first identifying where injury or ill health can occur, and then instituting measures that will make the occurrences less likely. Management should put in place organizational arrangements that are adapted to the size of the undertaking and the nature of its activities. Such arrangements should include the preparation of work procedures on the basis of job safety analysis. In this case, the person responsible should determine the safest, most effective way of performing a given task.

The place of occupational health and safety management

Occupational health and safety management should not be treated as a separate process, but be integrated into other workplace activities. Its various functions and procedures should be embedded in other management system and business processes in the enterprise, as well as within comparable structures in the community. For example, occupational health services in a small enterprise could be integrated with the primary health care provided within the community. This would be of benefit to workers and their families.

Performance measures

The ability to measure occupational health and safety performance over time is essential in order to verify that there is a continuous improvement in eliminating occupational injuries and illness. Employers should regularly verify the implementation of applicable standards on occupational health and safety, for instance by environmental monitoring, and should undertake systematic safety audits from time to time. Furthermore, they should keep records relating to occupational health and safety and the working environment, as specified by the competent authority. Such information might include records of all notifiable occupational accidents and injuries to health which arise in the course of or in connection with work, lists of authorizations and exemptions under laws or regulations relating to the supervision of the health of workers in the enterprise, and data concerning exposure to specified substances and agents.

A comprehensive evaluation system would include baseline evaluations, auditing, self-inspection and self-correction, incident investigation, medical surveillance, and management review activities.

PART III

OPERATIONAL MEASURES

6. Legislation, enforcement and collective agreements[1]

Appropriate legislation and regulations, together with adequate means of enforcement, are essential for the protection of workers' health and safety. Legislation is the very foundation of social order and justice; without it, or where it is not enforced, the door is wide open to all forms of abuse. Each country should therefore take such measures as may be necessary to protect workers' health and safety. This could be by enacting laws or regulations, or any other method consistent with national conditions and practice and in consultation with the representative organizations of employers and workers concerned. The law directly regulates certain components of working conditions and the work environment, including hours of work and occupational health and safety. There are also provisions relating to trade unions and collective bargaining machinery, which establish conditions for negotiations between employers and workers.

One of the greatest problems regarding labour legislation in many countries is its application in practice. It is therefore important for governments to take the necessary steps to ensure that there is an effective system of *labour inspection* to make certain that statutory requirements are met. This is often difficult because of a shortage of trained personnel. Another problem relates to the difficulty of deal-

[1] This chapter is based mainly on the Labour Inspection Convention, 1947 (No. 81), and Protocol, 1995, the Right to Organise and Collective Bargaining Convention, 1949 (No. 98), the Labour Inspection (Agriculture) Convention, 1969 (No. 129), the Occupational Safety and Health Convention (No. 155), and Recommendation (No. 164), 1981, and PIACT (op. cit.).

ing with new hazards, bearing in mind the speed at which technology is changing. In some cases such problems can be solved by employers and workers, through *collective bargaining*. These two complementary approaches are outlined below.

Labour inspection

The enforcement of legal provisions concerning occupational health and safety and the working environment should be secured by an adequate and appropriate system of inspection. The system should be guided by the provisions of the relevant ILO instruments,[2] without prejudice to the obligations of the countries that have ratified them.

As provided in Article 3(1) of the the Labour Inspection Convention, 1947 (No. 81), the functions of the system of labour inspection should be:

(a) to secure the enforcement of the legal provisions relating to conditions of work and the protection of workers while engaged in their work, such as provisions relating to hours, wages, safety, health and welfare, the employment of children and young persons, and other connected matters, in so far as such provisions are enforceable by labour inspectors;

(b) to supply technical information and advice to employers and workers concerning the most effective means of complying with the legal provisions;

(c) to bring to the notice of the competent authority defects or abuses not specifically covered by existing legal provisions.

For inspection to be taken seriously, labour legislation must be enforced systematically and forcefully. This may be a tall order in many countries because:

- legislation may not be sufficiently realistic;
- labour inspectors may have difficulty in imposing their authority;
- infrastructure facilities essential for inspection, such as adequate means of transport or communication, may not be available; and
- procedures may be lengthy and costly.

[2] See note 1, above.

Box 14. Cooperation between inspectors and workers

The Labour Inspection Convention, 1947 (No. 81) lays down standards for cooperation between inspectors and workers. In Article 5, the Convention states:

> The competent authority shall make appropriate arrangements to promote ... collaboration between officials of the labour inspectorate and employers and workers or their organisations.

In addition, Article 5 of the accompanying Recommendation (No. 81) states that representatives of the workers and the managements should be authorized to collaborate directly with officials of the labour inspectorate.

It is therefore imperative to broaden national labour inspection activities to involve employers and workers more actively (see box 14), and to make greater efforts in the field of training.

It should be stressed that any further duties which may be entrusted to labour inspectors should not be such as to interfere with the effective discharge of their primary duties or to prejudice in any way the authority and impartiality that are necessary to inspectors in their relations with employers and workers. Also, the need for the labour inspectorate staff to be well trained cannot be overemphasized.

In view of the crucial role of labour inspection regarding national occupational health and safety programmes, efforts must be made by the respective government authorities to strengthen the inspectorate. Depending on national approaches and circumstances, appropriate measures necessary to achieve the above objectives may include:

- improving the capacity to secure the enforcement of legal provisions;
- supplying technical information and advice;
- identifying new needs for action;
- an increase in the number of inspectors;

- improved training for inspectors in support of their enforcement and advisory role;
- the integration of separate inspection units or functions;
- the use of multidisciplinary inspection teams;
- closer cooperation between labour inspectors and employers, workers and their organizations;
- improved statistical reporting systems of occupational accidents and diseases and inclusion in the annual inspection report; and
- improved support facilities, institutions and other material arrangements.

The labour inspectorate must have an adequate and well-trained staff, be provided with adequate resources, have an effective presence at the workplace, and take decisive action by being severe, persuasive or explanatory, depending on the case.

It must be stated in conclusion that the conditions for an effective labour inspectorate, as discussed above, are very hard to attain in many countries of the world (see box 15). The reasons are not difficult to understand, and include scarce resources, especially in countries undergoing various programmes of economic reform, and the low priority given to occupational health and safety issues in the face of other competing demands. There is therefore very little justification for maintaining two parallel inspection systems, a practice that is still being observed in some countries. It is certainly more cost-effective to have an integrated system of inspection, whereby labour inspectors are also trained in health and safety issues. The mechanisms for achieving this process should be embodied in the national policy on occupational health and safety.

Collective bargaining

Since legislative processes are slow, collective agreements are particularly suitable for laying down requirements with respect to working conditions and the work environment in an enterprise. Collective bargaining is one of the most important and effective means of bringing about improvements in this field and should therefore be encouraged and promoted. It reflects the experience and interests of

Box 15. Some problems of labour inspection

A meeting of labour inspection experts dealing with child labour, organized by the ILO in 1999, made some broad observations of general interest to labour inspectors. The first problem identified was the lack of resources:

> In developing countries generally there was a great shortage of human and material resources to carry out the functions of labour inspection. There were perhaps genuine intentions to apply the law, but performance failed to measure up to these intentions. Posts existed but qualified inspectors could not be found and there were insufficient funds for training and purchasing equipment.

Another problem was interference from vested interests:

> Although Article 4 of Convention No. 81 was clear in stating that labour inspection should, if national law and practice so permitted, come under one central authority, some countries varied in the extent to which labour inspection was organized under a central, regional or local body. The further labour inspection was removed from this central authority, the greater the risk of involvement of vested interests in decisions affecting its independence. Pressure to change the manner of organizing had often occurred because of the perceived costs of running labour inspection without highlighting the benefits also in economic terms. This had been a particular issue in developing countries because of the regular requirement of many structural adjustment programmes to cut public expenditure and reduce public services more or less drastically. The impact on the independence and operation of labour inspection was therefore largely negative, with obvious consequences also for the ability of inspectors to meet the challenge of combating child labour.

Source: ILO: *Labour inspection and child labour*, Report of the Meeting of Experts on Labour Inspection and Child Labour, Geneva, 27 September-1 October 1999, pp. 5-6, para. 27; p. 8, para. 38.

the employers and workers concerned, as well as the economic, technical and social realities of particular trades, branches of activity or enterprises. As Article 4 of the Right to Organise and Collective Bargaining Convention, 1949 (No. 98) provides:

> Measures appropriate to national conditions shall be taken, where necessary, to encourage and promote the full development and utilisation of machinery for voluntary negotiation between employers and employers' organisations, with a view to the regulation of terms and conditions of employment by means of collective agreements.

Collective agreements are more flexible than legislation and are better adapted to local problems concerning working conditions and the environment, or the technical and economic problems of a given sector. They may also stipulate flexible procedures to resolve conflicts arising out of their application, as well as setting agreed time limits for their revision. With regard to occupational health and safety, collective agreements have often been used to bring about genuine progress and a tangible improvement in workers' conditions. This process is becoming increasingly common in small and medium-sized enterprises, especially when the improvement in occupational health and safety is linked not just to health issues, but also to increased productivity, higher-quality products and better morale among workers.

Another possible forum for the discussion of working conditions and the environment is that of works committees or other similar bodies. Depending on the country and their terms of reference, they may have different names and deal with different problems, either general (works committees, works councils) or specific (occupational health and safety committees). They may either be bilateral (composed of a variable number of workers' and management representatives) or consist of delegates elected by the workers or nominated by trade unions.

7. Occupational health surveillance[1]

Since the consequences of occupational hazards may not become apparent for many years, it is important to identify potential dangers early before they result in incurable diseases. The methods for identifying occupational hazards and the health problems associated with them can be broadly listed as environmental assessment, biological monitoring, medical surveillance and epidemiological approaches. Similar methods should be used to identify potential risks of accident. Some of the terminology associated with surveillance is explained in box 16.

Surveillance of the working environment

General framework

To ensure a healthy working environment there must be monitoring at the workplace. This involves systematic surveillance of the factors in the working environment and working practices which may affect workers' health, including sanitary installations, canteens and housing where these facilities are provided by the employer, as well as ensuring the quality of the working environment through compliance with safety and health standards.

---

[1] This chapter is based mainly on the Occupational Cancer Recommendation, 1974 (No. 147), the Occupational Safety and Health Convention, 1981 (No. 155), the Occupational Health Services Convention (No. 161), and Recommendation (No. 171), 1985, the Asbestos Convention, 1986 (No. 162), and the Chemicals Recommendation, 1990 (No. 177).

Everyone associated with the workplace – from the worker right through to the employer – should be actively involved in the surveillance of the working environment. Basic surveillance is carried out by simple observation and every worker, from shop floor to administration, should be trained to be able to identify those factors (potential or real) which may affect workers' health. Such training is necessary to enable the worker to report immediately to his or her direct supervisor any situation which can reasonably be thought to present an imminent and serious danger to life or health. In such a situation, the employer cannot require the worker to return to work until any necessary remedial action is taken.

Box 16. Surveillance, work and health

Information about conditions in the working environment and the health of workers – which is necessary for planning, implementing and evaluating occupational health programmes and policies – is gathered through ongoing, systematic surveillance. Different types of surveillance address the various aspects of work and health. Some activities focus principally on the health of workers themselves, while others explore the various factors in the work environment that may have negative impacts on health. Whatever the approach taken, researchers must meet minimum requirements with regard to workers' sensitive health data.

Workers' health surveillance entails procedures for the assessment of workers' health by means of detection and identification of any abnormalities. Such procedures may include biological monitoring, medical examinations, questionnaires, radiological examinations and reviews of workers' health records, among others.

Working environment surveillance involves identification and assessment of environmental factors that may affect workers' health, such as the state of occupational hygiene and sanitation, organization of work, personal protective equipment and control systems, and workers' exposure to hazardous substances. Such surveillance may focus on accident and disease prevention, ergonomics, occupational hygiene, organization of work and psychosocial factors, among others.

For more information, see ILO: *Technical and ethical guidelines for workers' health surveillance*, Occupational Safety and Health Series, No. 72 (Geneva, 1999).

Simple observation (a walk-through survey) of work processes and the working environment is the first step in any surveillance. Such observation may be sufficient in some cases to detect a lack of adequate control measures and of exposure of workers to risk. The evaluation based on this type of observation may justify the recommendation of control measures without the need for any more sophisticated determination of the level of exposure. Visits to the workplace and walk-through observation are also necessary to provide an assurance that no deterioration has occurred at workplaces initially evaluated as satisfactory.

Information from surveillance of the working environment should be combined with other data, such as epidemiological research or exposure limits, to assess occupational health risks. For a definition of the main concepts involved in risk assessment, see box 17.

Box 17. Risk assessment

Risk assessment is an increasingly popular tool for analysing workplace hazards. The method rests upon clear definitions of the two terms: HAZARD and RISK.

HAZARD is defined as:

the *potential to cause harm* – which can include substances or machines, methods of work or other aspects of organization.

RISK is defined as:

the *likelihood* that the harm from a particular hazard is realized.

Further important definitions are:

Likelihood of occurrence (probability):

- Low: remote or unlikely to occur.
- Medium: will occur in time if no preventive action taken.
- High: likely to occur immediately or in near future.

Consequence (severity):

- Low: may cause minor injury/illness – no lost time.
- Medium: may cause lost time injury/illness.
- High: may cause serious or fatal injury/illness.

Using these definitions, a risk matrix can be constructed.

For example, when there is a high likelihood that workers will be exposed to a hazard, and the consequences are high, then that work, process, chemical, and so on, would have a high "score" and urgent action should be taken.

An approach which is gaining widespread support is the *precautionary principle*. Simply stated, it is the need to foresee and forestall damaging human activities before science delivers irrefutable proof that there is a problem. We have seen many examples where concerns have been raised about a substance or process, to be told that there is no proof that it is harmful. By the time proof is established, hundreds, if not thousands, of people may have died, or suffered irreversible damage to their health.

When an activity raises threats of harm to human health or the environment, precautionary measures should be taken even if some cause and effect relationships are not fully established scientifically. This is articulated in four clauses:

- people have a duty to take anticipatory action to prevent harm;
- the burden of proof of harmlessness of a new technology, process, activity or chemical lies with the activity's proponents, not with the general public;
- before using a new technology, process or chemical, or starting a new activity, people have an obligation to examine a full range of alternatives, including the alternative of doing nothing; and
- decisions applying the precautionary principle must be open, informed and democratic, and must include affected parties.

The precautionary principle has now been incorporated into some international treaties and some national laws (e.g. Sweden's chemical laws and the laws of some American states.)

Monitoring of exposure

There may be special health hazards which require particular monitoring. Surveillance programmes should therefore include the monitoring of workers' exposure to such hazards, when necessary. The main objectives of such monitoring are to:

- identify real hazards;
- determine the level of workers' exposure to harmful agents;
- prove compliance with regulatory requirements;

- assess the need for control measures; and
- ensure the efficiency of control measures in use.

The above objectives can be achieved by carrying out occupational health surveys in addition to routine monitoring programmes. Occupational health surveys are defined as investigations of environmental conditions in the workplace, conducted primarily to determine the nature and extent of any condition that may adversely affect the well-being of persons engaged there. Such surveys are necessary to develop the engineering and medical control measures needed to eliminate or avoid harmful situations.

There are two types of occupational health surveys:

- the walk-through survey, which is made for the purpose of selecting any locations in the plant where exposure to hazards are later to be evaluated by analytical studies in order to determine whether additional control is necessary; and
- the comprehensive occupational health survey, which involves the use of sophisticated monitoring equipment and entails proper planning and execution.

In a situation where workers are exposed to hazardous substances, for example airborne toxic chemicals, the employer should:

- limit exposure to such substances so as to protect the health of workers; and
- assess, monitor and record, as necessary, the concentration of substances at the workplace.

The monitoring of exposure should be carried out in accordance with the requirements of the competent authority. The competent authority should further establish the criteria for determining the degree of exposure to the substances or agents in question, and where appropriate should specify levels as indicators for surveillance of the working environment in accordance with the technical preventive measures in place, as for example occupational cancer. Monitoring should be carried out and assessed by trained and experienced people, in accordance with recognized and scientifically accepted methods.

The monitoring strategy should assess both the current situation and the possible effect of technological changes or control measures, for example on the concentration of air pollutants, and be conducted with a number of specific aims in view (box 18).

Box 18. Aims of a monitoring strategy for air pollutants

Monitoring of exposure should ensure that:

- specific operations where exposure may occur are identified and levels of exposure are quantified;
- exposure to air pollutants does not exceed exposure limits set or approved by the competent authority;
- effective preventive measures are implemented for all applications and in all jobs;
- any changes in manufacturing, use or work practices do not lead to increased exposure to air pollutants; and
- supplementary preventive measures are developed as necessary.

Exposure limits

One of the responsibilities of the competent authority is to establish the criteria for determining the degree of exposure to hazardous substances or agents, and where appropriate to specify levels as indicators for surveillance of the working environment, with a view to implementing the technical preventive measures required. Furthermore, the competent authority is required to prescribe limits for the exposure of workers to hazardous substances, for example asbestos. Such exposure limits or criteria for determining the degree of exposure are to be fixed, periodically reviewed, and updated in the light of technological progress and advances in technological and scientific knowledge.

The exposure limits are usually expressed as time-weighted, whole-shift concentrations and, where necessary, short-term peak concentrations. In practice, the concentration of air pollutants cannot be measured at all workstations and at all times. A limited number of representative air samples are usually taken in order to estimate the average concentration of the pollutants in the workplace. This concentration can then be compared with the exposure limit. The sampling site and duration should be selected so as to ensure the representativeness of the result. The sampling should be carried out at fixed sites (area sampling) or at the breathing zone of the worker (personal sampling). Unless self-reading instruments are used, the samples will have to be analysed later by appropriate methods.

Exposure limits are not a simple mechanism. There are a number of reservations which must be borne in mind:

- Exposure limits are based implicitly on male North American workers, who have a greater body weight than most women workers and most male Asian workers, for example. The latter categories should therefore be exposed to "doses" lower than the United States-based standards, often used as the norm.
- They do not represent a sharp dividing line between "safe" and "hazardous" levels.
- The absence of a substance from tables and lists produced by a competent authority is sometimes taken as evidence that it is safe. For many substances, no limit is set.
- Limits are based on the assumption that exposure is limited to one substance only. However, in many workplaces there will be a variety of chemicals, forming a "cocktail" which can represent a greater danger.
- Other factors, such as high temperature and humidity, long hours of work and ultra-violet radiation may increase the toxic response to a substance.

The surveillance of the working environment should be carried out in liaison with other technical services in the enterprise, in cooperation with the workers concerned and with their representatives in the enterprise or the safety and health committee, where they exist (see box 19). Occupational health monitoring services should be able to call on sufficient technical expertise in relevant fields. The evaluation of pollution levels and workers' exposure requires specialist knowledge. Such evaluation should therefore be carried out by, or in close cooperation with, an experienced industrial hygienist.

Box 19. Responsibilities of staff involved in the surveillance of the working environment

Staff involved in the surveillance of the working environment are responsible for:

- conducting surveys of the working environment;
- interpreting the data gathered during the survey;
- keeping records;
- preparing appropriate control measures;
- preparing adequate warnings;
- suggesting precautions where dangers exist;
- advising management on industrial hygiene;
- educating workers and the community at large on basic occupational safety and health;
- conducting epidemiological studies to uncover the presence of occupation-related illness and injury.

Record-keeping

The results of workplace monitoring should be collected in a standardized way. Employers should keep the records of the monitoring of exposure for the period determined by the competent authority. This is to enable the assessment of any possible relation between later health impairment and exposure. For example, in cases of exposure to silica, coal, asbestos or carcinogenic substances, it may be necessary to keep records for several decades. Arrangements should also be made by the competent authority to conserve the records in an archive, so that they remain available even though an enterprise may close down. Records should include all relevant data, such as details of the site, product, manufacturer and methods of use, including the availability and wearing of personal protective clothing or equipment. Workers and their representatives and the competent authority should have access to the monitoring records as in the case of chemicals, for example.

Surveillance of workers' health

General framework

The surveillance of workers' health entails medical examinations of workers to ensure that their state of health is compatible with their job assignment and that their occupational exposure does not have any detrimental effects on their health. Health examinations help to identify conditions which may make a worker more susceptible to the effects of hazardous agents or detect early signs of health impairment caused by them. Their main purpose is primary prevention of work-related injuries and diseases. Surveillance should be carried out in consultation with the workers or their representatives, and should not result in any loss of earnings for them. Furthermore, medical examinations should be free of charge and, as far as possible, should take place during working hours.

Workers' health surveillance at national, industry and enterprise levels should be organized so as to take into account several factors, including:

- the need for a thorough investigation of all work-related factors;
- the nature of occupational hazards and risks in the workplace which may affect workers' health;
- the health requirements of the working population;
- the relevant laws and regulations and the available resources;
- the awareness of workers and employers of the functions and purposes of such surveillance; and
- the fact that surveillance is not a substitute for monitoring and control of the working environment.

Medical examinations, health assessments and biological tests

Workers who are or have been exposed to occupational hazards, such as asbestos, should be provided with such medical examinations as are necessary to supervise their health in relation to those occupational hazards, and to diagnose occupational diseases caused by exposure to such hazards.

Surveillance of workers' health in the form of medical screening or periodic medical examinations often leads to the identification of occupational hazards or diseases. It has been shown that special prescriptive surveys to detect ill health among the working population generally prove more rewarding in terms of avoiding or controlling hazards than a series of medical tests performed at a later stage to identify or confirm suspected occupational disease. Cases of occupational disease often remain "latent" (silent) among the labour force. The workers adapt to a slowly developing condition, and are often unwilling to report illness that may result in the loss of their jobs.

Health examinations of workers frequently reveal the existence of health hazards in the workplace, and in such cases the necessary environmental evaluation and control measures must be implemented.

The importance of workers' health surveillance is clearly stated in paragraph 11 of ILO Recommendation No. 171, which provides as follows:

> ... Surveillance of the workers' health should include, in the cases and under the conditions specified by the competent authority, all assessments necessary to protect the health of the workers, which may include:
>
> (a) health assessment of workers before their assignment to specific tasks which may involve a danger to their health or that of others;
>
> (b) health assessment at periodic intervals during employment which involves exposure to a particular hazard to health;
>
> (c) health assessment on resumption of work after a prolonged absence for health reasons for the purpose of determining its possible occupational causes, of recommending appropriate action to protect the workers and of determining the worker's suitability for the job and needs for reassignment and rehabilitation;
>
> (d) health assessment on and after the termination of assignments involving hazards which might cause or contribute to future health impairment.

Pre-assignment medical examinations are carried out before the job placement of workers or their assignment to specific tasks which may involve a danger to their health or that of others. The purpose of this health assessment is to determine in what capacity the prospective employee can be utilized most efficiently without detriment to himself or herself or to fellow workers. The scope of pre-assignment medical examination is influenced by such factors as the nature and location of the industry, as well as by the availability of the services of physicians and nurses. Regardless of the size of the enterprise, it is advisable to conduct such examinations for all prospective employees. In the case of young persons, such pre-assignment medical examinations are prescribed by specific ILO Conventions.

The pre-assignment medical examination provides the necessary clinical information and laboratory data on the worker's health status at the moment of entering employment. It is also important with regard to the worker's subsequent occupational history, as it provides a baseline for the subsequent evaluation of any changes in health status that may occur later on. The results of pre-assignment medical examinations should be used to help place workers in jobs which are compatible with the status of their health, and not to screen out workers. In some cases, prospective employees who are found to be HIV-positive may be refused employment on the basis of their health status, or those already in employment may be summarily dismissed. These practices should not be condoned.

Periodic health evaluations are performed at appropriate intervals during employment to determine whether the worker's health remains compatible with his or her job assignment and to detect any evidence of ill health attributed to employment. Their objectives include:

- identifying as early as possible any adverse health effects caused by work practices or exposure to potential hazards; and
- detecting possible hazards.

Changes in the body organs and systems affected by harmful agents can be detected during the periodic medical examination, usually performed after the worker has been employed long enough to have been exposed to any such hazards in the workplace. The worker may be physically fit, showing no signs of impairment and unaware of the fact that the substances he or she works with daily are slowly poisoning his or her system. The nature of the exposure and the expected biological response will determine the frequency with which the periodical medical examination is conducted. It could be as frequent as one to three months, or it could be carried out at yearly intervals.

Return-to-work health assessment is required to determine the worker's suitability for resumption of duty after a prolonged absence for health reasons. Such an assessment might recommend appropriate actions to protect the worker against future exposure and determine the need for reassignment or special rehabilitation. A similar assessment is performed on a worker who changes job, with a view to certifying him or her fit for the new duties.

Post-assignment health examinations are conducted after the termination of assignments involving hazards which could cause or contribute to future health impairment. The purpose is to make a final evaluation of workers' health and compare it with previous medical examinations to see whether the job assignments have affected their health.

In certain hazardous occupations, the competent authority should ensure that provision is made, in accordance with national law and practice, for appropriate medical examinations to continue to be available to workers after the termination of their assignment.

At the conclusion of a prescribed health assessment, workers should be informed in a clear and appropriate manner, by the attending physician, of the results of their medical examinations and receive individual advice concerning their health in relation to their work. When such reports are communicated to the employer, they should be devoid of any information of a medical nature. They should simply contain a conclusion about the fitness of the examined person for the proposed or held assignment and specify the kinds of jobs and conditions of work which are medically contraindicated either temporarily or permanently.

When continued assignment to work involving exposure to hazardous substances is found to be medically inadvisable, every effort, consistent with national conditions and practice, should be made to provide the workers concerned with other means of maintaining an income. Furthermore, national laws or regulations should provide for the compensation of workers who contract a disease or develop a functional impairment related to occupational exposure, in accordance with the Employment Injury Benefits Convention, 1964 (No.121).

It must be mentioned that there are limitations to medical examinations especially in the developing countries where generally the provision and coverage of health services is poor and the doctor to patient ratio is very high. The heavy workload and other limitations often affect the thoroughness of medical examinations.

In the case of exposure to specific occupational hazards, special tests are needed. These should be carried out in addition to the health assessments described above. The surveillance of workers' health should thus include, where appropriate, any other examinations and investigations which may be necessary to detect exposure levels and early biological effects and responses.

The analysis of biological samples obtained from the exposed workers is one of the most useful means of assessing occupational exposure to a harmful material. This analysis may provide an indication of the amount of substance that has accumulated or is stored in the body, the amount circulating in the blood, or the amount being excreted. There are several valid and generally accepted methods of biological monitoring which allow for the early detection of the effects on workers' health of exposure to specific occupational hazards. These can be used to identify workers who need a detailed medical examination, subject to the individual worker's consent. Urine, blood and saliva are the usual body fluids examined for evidence of past exposure to toxic (harmful) agents. Lead concentrations in the urine or blood have long been used as indices of lead exposure.

Priority should be given to environmental criteria over biological criteria in setting exposure limits, even though biological monitoring has certain advantages over environmental sampling. In biological monitoring, substances being absorbed through the skin and gastrointestinal tract (stomach) are accounted for, and the effects of added stress (such as increased workload resulting in a higher respiration rate with increased intake of the air contaminant) will be reflected in the analytical results. Furthermore, the total exposure (both on and off the job) to harmful materials will be accounted for. Biological monitoring should not, however, be a substitute for surveillance of the working environment and the assessment of individual exposures. In assessing the significance of the results of biological monitoring, values commonly found in the general public should be taken into account.

Sickness absence monitoring

The importance of keeping a record of sickness absenteeism is well recognized in various countries. Monitoring sickness absence can help identify whether there is any relation between the reasons for ill health or absence and any health hazards which may be present at the workplace. Occupational health professionals should not, however, be required by the employer to verify the reasons for absence from work. Their role is rather to provide advice on the health status of the workforce in the enterprise and on medical problems which affect attendance and fitness for work. Occupational health professionals should not become involved in the administrative management and control of sickness absence, but it is acceptable for them to advise on medical aspects of sickness cases, provided that medical confidentiality is respected.

Reporting of occupational accidents, injuries and diseases

One of the tasks of the competent authority is to ensure the establishment and application of procedures for the notification of occupational accidents and diseases by employers and, when appropriate, insurance institutions and others directly concerned, as well as the production of annual statistics on occupational accidents and diseases. Consequently, national laws or regulations in many countries provide for:

- the reporting of occupational accidents and diseases to the competent authority within a prescribed time;
- standard procedures for reporting and investigating fatal and serious accidents as well as dangerous occurrences; and
- the compilation and publication of statistics on accidents, occupational diseases and dangerous occurrences.

This compulsory reporting is usually carried out within the framework of occupational disease and injury prevention, compensation or benefit programmes. There are also voluntary occupational injury and disease reporting systems. In any case, the competent authority is responsible for developing a system of notification of occupational diseases, in the case of asbestos for example. It must be acknowledged that occupational diseases, compared to occupational accidents, are usually not well recorded since the factors of recognition are different between countries. Member States could use the ILO code of practice *Recording and notification of occupational accidents and diseases* (Geneva, 1996) as a basis for developing their own systems.

Whatever the system developed, it is the responsibility of the employer to present a detailed report to the competent authority within a fixed period. After a major accident, for example, the employer must submit a report containing an analysis of the causes of the accident and describing its immediate on-site consequences, as well as indicating any action taken to mitigate its effects. It is equally the responsibility of the employer to keep records of relevant occupational accidents and diseases. In this respect, it is worth pointing out that good record-keeping is beneficial to the company in many ways (see box 20).

Box 20. Some benefits of good record-keeping

- The company is able to assess the economic impact of accidents in terms of production time lost, damage to machinery or raw materials, product liability and increased premiums paid to the workers' compensation insurance fund.
- Having assessed the economic consequences and the common type of accidents at its workplace, the company can identify "high-risk" occupations and processes and devise better accident-prevention strategies in future to minimize or eliminate accidents at work.
- An accident-free workplace enhances worker morale, improves worker/management relationships, and leads to increased productivity and fewer industrial disputes.
- The public image of a company improves if there are minimal or no accidents at all, and this will have a positive effect on the sale of its products.
- If a proper register of accidents is kept, the company will have nothing to fear when inspectors visit to inspect or investigate accidents.

In many countries, lists of notifiable occupational diseases have been established by statute. The records of notified diseases give administrators some idea of the extent and types of occupational pathology. This presupposes that medical practitioners are well informed in such diagnoses and are prepared to cooperate, which unfortunately is not always the case. Workers' compensation schemes operated by ministries of labour also have lists covering occupational injuries that are subject to compensation.

Where an occupational disease has been detected through the surveillance of the worker's health, it should be notified to the competent authority, in accordance with national law and practice. The employer, workers and workers' representatives should be informed that this notification has been carried out. Specifically, the labour inspectorate, where it exists, should be notified of industrial accidents and occupational diseases in the cases and in the manner prescribed by national laws and regulations.

Ethical and legal issues

The surveillance of workers' health should be based not only on sound technical practice, but on sound ethical practice as well. In this regard, a number of conditions must be met (box 21) and workers' rights respected. In particular, workers subject to health monitoring and surveillance should have:

- the right to confidentiality of personal and medical information;
- the right to full and detailed explanations of the purposes and results of the monitoring and surveillance; and
- the right to refuse invasive medical procedures which infringe on their corporal integrity.

Box 21. Conditions governing workers' health surveillance

- Provisions must be adopted to protect the privacy of workers and to ensure that health surveillance is not used for discriminatory purposes or in any other manner prejudicial to workers' interests.
- Each person who works in an occupational health service should be required to observe professional secrecy as regards both medical and technical information which may come to his or her knowledge in connection with the activities of the service, subject to such exceptions as may be provided for by national laws or regulations.
- Occupational health services should record data on workers' health in personal confidential health files which should also contain information on jobs held by workers, on exposure to occupational hazards involved in their work, and on the results of any assessments of workers' exposure to these hazards.
- Although the competent authority may have access to data resulting from the surveillance of the working environment, such data may only be communicated to others with the agreement of the employer and the workers or their representatives in the enterprise or the safety and health committee.
- Personal data relating to health assessments may be communicated to others only with the informed consent of the worker concerned.

8. Occupational health services[1]

General considerations

Occupational health services are defined as services entrusted with essentially preventive functions. As provided in the Occupational Health Services Convention, 1985 (No. 161), they are responsible for advising the employer, the workers and their representatives in the workplace on:

(i) the requirements for establishing and maintaining a safe and healthy working environment which will facilitate optimal physical and mental health in relation to work;

(ii) the adaptation of work to the capabilities of workers in the light of their state of physical and mental health.[2]

It is desirable that some sort of occupational health services be established in every country. This may be done either by laws or regulations, or by collective agreements, or as otherwise agreed upon by the employers and workers concerned, or in any other manner approved by the competent authority after consultation with the representative organizations of employers and workers concerned.

---

[1] This chapter is based mainly on the Occupational Safety and Health Recommendation, 1981 (No. 164), the Occupational Health Services Convention (No. 161), and Recommendation (No. 171), 1985, the Safety and Health in Construction Convention (No. 167), and Recommendation (No. 175), 1988, the Safety and Health in Mines Recommendation, 1995 (No. 183), and PIACT (op. cit.).

[2] Article 1(a).

The coverage of workers by occupational health services varies widely, ranging from 5-10 per cent at best in the developing world up to 90 per cent in industrialized countries, especially those in Western Europe. There is therefore a universal need to increase worker coverage throughout the world.

Ideally, each country should progressively develop occupational health services for all workers, including those in the public sector and members of production cooperatives, in all branches of economic activity and in all enterprises. The occupational health services provided should be adequate and appropriate to the specific health risks of the enterprises. These services should also include the necessary measures to protect self-employed persons and informal sector operators. To that end, plans should be drawn up to effect such measures and to evaluate progress made towards their implementation.

The main concepts involved in occupational health are defined in box 22 (pp. 84-85).

Organization

Occupational health services can be organized as a service for a single enterprise or as a service common to a number of enterprises, depending on which type is more appropriate in terms of national conditions and practice. Similarly, these services may be organized by:

- the enterprise or group of enterprises concerned;
- the public authorities or official services;
- social security institutions;
- any other bodies authorized by the competent authority; or
- a combination of any of the above.

In the absence of a specific occupational health service, the competent authority may, as an interim measure, designate an appropriate existing service, for instance a local medical service, to act as an occupational health service.

Thus, in an enterprise where establishing an occupational health service or having access to such a service is impracticable, the competent authority should – after consulting the employers' and workers' representatives in the workplace or the safety and health committee – make provisional arrangements with a local medical service to:

- carry out the health examinations prescribed by national laws or regulations;
- ensure that first aid and emergency treatment are properly organized; and
- provide surveillance of the environmental health conditions in the workplace.

Occupational health services should be made up of multidisciplinary teams whose composition should be determined by the nature of the duties performed. Each team should have sufficient technical personnel with specialized training and experience in such fields as occupational medicine, occupational hygiene, ergonomics and occupational health nursing. The staff of occupational health services should, as far as possible, keep themselves up to date with progress in the scientific and technical knowledge necessary to perform their duties and should be given the opportunity to do so without loss of earnings. In addition, occupational health services should have the necessary administrative personnel to ensure their smooth operation. The staff of occupational health services must enjoy full professional independence from employers, workers and their representatives in relation to the functions of occupational health services.

Functions

Basically, occupational health services aim to protect and promote the health of workers, improve working conditions and the working environment, and maintain the health of the enterprise as a whole (see box 23 for details).

Box 22. Concepts in occupational health

Numerous concepts dealing with the interaction between work and health underpin occupational health and safety programmes.

In relation to work, "health" indicates not merely the absence of disease or infirmity; it also includes the physical and mental elements affecting health which are directly related to safety and hygiene at work.

Occupational health practice is a broad concept, and includes *occupational health services*, which are defined in Article 1(a) of the Occupational Health Services Convention, 1985 (No. 161). It involves activities for the protection and promotion of workers' health and for the improvement of working conditions and environment carried out by occupational health and safety professionals as well as other specialists, both within the enterprise and without, as well as workers' and employers' representatives and the competent authorities. Such multisectoral and multidisciplinary participation demands a highly developed and coordinated system in the workplace. Administrative, operative and organizational systems must be in place to conduct occupational health practice successfully.

Occupational health care is another broad concept; it encompasses all the people and programmes directly or indirectly involved in making the work environment healthy and safe. It includes practical, enterprise-level efforts aimed at achieving adequate occupational health, such as preventive health care, health promotion, curative health care, first aid, rehabilitation and compensation as well as strategies for prompt recovery and return to work. Primary health care can also be considered as playing a part.

Occupational hygiene (sometimes also known as industrial hygiene) is the art and science of protecting workers' health through control of the work environment; it encompasses recognition and evaluation of those factors that may cause illness, lack of well-being or discomfort among workers or the community. As a component of occupational health and safety efforts to improve working conditions, occupational hygiene focuses on three major areas:

- recognition of the interrelation between environment and industry;
- factors of the working environment that may impair health and well-being;
- the formation of recommendations for the alleviation of such problems.

Occupational safety and health management comprises the activities designed to facilitate the coordination and collaboration of workers' and employers' representatives in the promotion of occupational health and safety in the workplace. The concept defines rights, roles and responsibilities regarding the identification of hazards and risks and the implementation of control or preventive measures.

For further discussion of occupational health practices, services and related concepts and terminology, see Igor A. Fedotov, Marianne Saux and Jorma Rantanen (eds.): "Occupational health services", in *Encyclopaedia of occupational health and safety*, op. cit., Vol. I, pp. 16.1-62.

Box 23. Functions of an occupational health service

The main functions of an occupational health service are to:

- identify and assess the risks from health hazards in the workplace;
- watch for factors in the work environment and working practices that may affect workers' health, such as sanitary installations, canteens and housing provided by the employer;
- advise on work planning and organization, including workplace design and the choice, maintenance and condition of machinery, and other equipment and substances used in work;
- participate in the development of programmes for the improvement of work practices;
- collaborate in testing new equipment and evaluating its health aspects;
- advise on occupational health, safety and hygiene, and on ergonomics and protective equipment;
- monitor workers' health in relation to work;
- try to make sure that work is adapted to the worker;
- contribute to vocational rehabilitation;
- collaborate in providing training and education in occupational health and hygiene, and ergonomics;
- organize first aid and emergency treatment; and
- participate in the analysis of occupational accidents and occupational diseases.

In order that they may perform their functions efficiently, occupational health services should:

- have free access to all workplaces and to the ancillary installations of the enterprise;
- be able to inspect the workplaces at appropriate intervals in cooperation, where necessary, with other services of the enterprise;
- have access to information concerning the processes, performance standards and substances used or the use of which is contemplated;
- be authorized to request the competent authorities to ensure compliance with occupational health and safety standards; and
- be authorized to undertake, or to request that approved technical bodies undertake:
  (i) surveys and investigations on potential occupational health hazards, for example by sampling and analysis of the atmosphere of workplaces, of the products and substances used, or of any other material suspected of being harmful; and
  (ii) the assessment of harmful physical agents.

Within the framework of their responsibility for their employees' health and safety, employers or management should take all necessary measures to facilitate the activities of occupational health services. Equally, workers and their organizations should provide support to occupational health service functions. Furthermore, in cases where occupational health services are established by national laws or regulations, the manner of financing these services should also be determined.

Primary health care approach

Programmes introduced in a number of countries to increase worker coverage have demonstrated that it is possible to substantially improve the availability of occupational health services in a relatively short time and at a reasonable cost by adopting a primary health care approach. Such an approach is particularly appropriate for developing countries as it has been found to improve both the workers' access to the services and the cost effectiveness of the services provided. Primary health care, in the community, by community-based doctors, nurses and other medical staff, can reach more people, often at lower cost than centralized hospital provision.

Bearing in mind that workers are part of the community at large, and taking into account the organization of preventive medicine at the national level, occupational health services might, where possible and appropriate:

- carry out immunizations in respect of biological hazards in the working environment;
- take part in campaigns for the protection of health; and
- collaborate with the health authorities within the framework of public health programmes.

First aid

In the context of occupational health and safety, first aid means the immediate measures taken at the site of an accident by a person who may not be a physician but who is trained in first aid, has access to the necessary equipment and supplies, and knows what must be done to ensure that professional medical care will follow his or her intervention.

When a serious accident happens, the first few minutes may be decisive in terms of lives being saved or injuries avoided. Therefore, taking national law and practice into account, occupational health services in enterprises should:

- provide first aid and emergency treatment in cases of accident or indisposition of workers at the workplace; and
- collaborate in the organization of first aid.

It is the employer's responsibility to ensure that first aid is available at all times. This implies that the employer has a responsibility to ensure that trained personnel are available at all times. It is up to the occupational health service to ensure the training and regular retraining of first-aid personnel. Indeed, on a broader scale, occupational health services should ensure that all workers who contribute to occupational safety and health are trained progressively and continuously.

The manner in which first-aid facilities and personnel are to be provided should be prescribed by national laws or regulations which should be drawn up after consulting the competent authority and the most representative organizations of employers and workers concerned.

Curative health services and rehabilitation

Although occupational health services essentially focus on prevention, they may also engage in other health activities, where the local organization of health care or the distance from general medical clinics makes such extended activities appropriate. These may include curative medical care for workers and their families, provided that the activities are authorized by the competent authority in consultation with the most representative organizations of employers and workers.

Measures should be taken to encourage and promote programmes or systems aimed at the rehabilitation and reintegration, where possible, of workers unable to undertake their normal duties because of occupational injury or illness. The provision of such care, where appropriate, should not involve any cost to the worker and should be free of any discrimination or retaliation whatsoever.

Special occupational health needs

Some workers have special occupational health needs for a variety of reasons, including age, physiological condition, social conditions and barriers to communication. The special needs of such workers should be met on an individual basis with due concern to protecting their health at work, making sure that there is no possibility of discrimination.

One category of workers with special occupational health needs comprises pregnant women and working mothers. The assessment of risks at work, and the preventive and control strategies prescribed to control risks should take account of these special needs, and arrangements should be made to avoid harm. Where reproductive health hazards and risks have been identified, employers should take appropriate measures. This is especially important during health-risk periods such as pregnancy and breast-feeding. These measures might include training and special technical and organizational measures, in particular the right to appropriate alternative work, without any loss of salary.

Disabled workers are another category of workers with special needs. In order to give effect to occupational safety and health policy, the competent authority should provide appropriate measures for these workers. Similarly, measures should also be taken to promote programmes or systems for the rehabilitation and reintegration of workers who have sustained occupational injury or illness.

Cooperation and coordination

Occupational health services should carry out their functions in cooperation with the other services in the enterprise. Measures should be taken to ensure adequate cooperation and coordination between occupational health services and, as appropriate, other bodies concerned with the provision of health services. It is recommended that national laws and practice be adapted to these requirements to ensure progress in the field of occupational health and safety.

Occupational health services should cooperate with the other services concerned in the establishment of emergency plans for action in the case of major accidents. Where necessary, they should also have contacts with external services and bodies dealing with questions of health, hygiene, safety, vocational rehabilitation, retraining and reassignment, working conditions and the welfare of workers, inspection, as well as with the national body designated to take part in the International Occupational Safety and Health Hazard Alert System (see box 24) set up within the framework of the International Labour Organization.

The occupational health service of a national or multinational enterprise with more than one establishment should provide the highest standard of services, without discrimination, to the workers in all its establishments, regardless of the place or country in which they are situated.

Box 24. The ILO and occupational health and safety information

The ILO's activities in the realm of occupational health and safety include dissemination of practical information. In particular, three major complementary tools provide the organization's constituents (governments, employers and workers) with information on occupational safety and health:

- the four volumes of the *Encyclopaedia of occupational health and safety*, 4th edition, published in 1998;
- the International Occupational Safety and Health Information Centre (CIS), which produces a database of occupational health and safety literature; and
- the International Occupational Safety and Health Hazard Alert System, which disseminates rapidly, through a worldwide network, scientific and technical information on newly discovered or suspected occupational health hazards.

The ILO publishes numerous other works on occupational health and safety and organizes scientific and technical meetings, congresses and symposia.

In addition, codes of practice offer useful guidelines to workers' and employers' organizations. They provide information for those charged with drawing up occupational health and safety regulations, programmes and policies. The ILO Governing Body approves the publication of codes of practice, which are not intended to replace national laws, regulations or accepted standards. The ILO has issued more than 20 codes of practice covering specific risks or potentially hazardous substances and various branches of economic activity (see Annex V).

Research

Within the limits of their resources, occupational health services should contribute to research by participating in studies or inquiries in the enterprise or in the relevant branch of economic activity. Such research should be subject to prior consultation with employers' and workers' representatives and might, for example, aim to collect data for epidemiological purposes or for orienting the activities of the occupational health service. The results of measurements carried out in the working environment and the assessments of workers' health may also be used for research purposes, subject to the agreement of the employer and the workers or their representatives in the enterprise or the safety and health committee. Most importantly, the privacy of the workers must be protected.

There is a need for action-oriented research programmes, in particular to:

- provide accurate statistics on the incidence of occupational accidents and diseases and on their causes;
- identify the hazards associated with all forms of new technology, including chemical substances;
- describe and analyse the working conditions of workers in poorly protected occupations and sectors; and
- investigate relationships between working conditions, occupational safety and health, and productivity, including the impact of improved conditions on employment and economic growth.

Where guidelines for research programmes are developed on a tripartite basis, the links between research and action are likely to be strengthened. Mechanisms to promote such tripartite collaboration should be put in place.

9. Preventive and protective measures [1]

General considerations

The incidence of accidents and work-related diseases and injuries in most occupational sectors is still regrettably high; there is therefore an urgent need for preventive and protective measures to be instituted at workplaces in order to guarantee the safety and health of workers. Occupational accidents and diseases not only cause great pain, suffering or death to victims, but also threaten the lives of other workers and their dependants. Occupational accidents and diseases also result in:

- loss of skilled and unskilled but experienced labour;
- material loss, i.e. damage to machinery and equipment well as spoiled products; and
- high operational costs through medical care, payment of compensation, repairing or replacing damaged machinery and equipment.

Workplace health and safety programmes should aim at eliminating the unsafe or unhealthy working conditions and dangerous acts which account for nearly all occupational accidents and diseases. Pre-

[1] This chapter is based mainly on the Working Environment (Air Pollution, Noise and Vibration) Recommendation, 1977 (No. 156), the Occupational Safety and Health Recommendation, 1981 (No. 164), the Asbestos Convention (No. 162), and Recommendation (No. 172), 1986, the Safety and Health in Construction Convention, 1988 (No. 167), the Chemicals Recommendation, 1990 (No. 177), the Safety and Health in Mines Convention, 1995 (No. 176), and PIACT (op. cit.).

vention by eliminating or reducing the sources of potential risks and the causes that trigger hazards can be achieved in a number of ways: engineering control, design of safe work systems to minimize risks, substituting safer materials for hazardous substances, administrative or organizational methods, and use of personal protective equipment.

Occupational health problems arise to a great extent from hazardous factors in the working environment. Since most hazardous conditions at work are in principle preventable, efforts should be concentrated on primary prevention at the workplace, as this offers the most cost-effective strategy for their elimination and control. Consideration should therefore be given to criteria and actions for the planning and design of healthy and safe workplaces in order to establish working environments that are conducive to physical, psychological and social well-being. This means taking all reasonable precautions to avoid occupational diseases and injuries.

The prevention of occupational hazards depends on the nature of the various causal agents, their mode of action and the severity of the risk. In prescribing measures for the prevention and control of hazards in the working environment, the competent authority should take into consideration the most recent ILO codes of practice or guides and the conclusions of relevant meetings of experts convened by the ILO, as well as information from other competent bodies. In taking preventive and protective measures the employer should assess the risks and deal with them in the order of priority set out in box 12. In situations where workers are exposed to physical, chemical or biological hazards, there are a number of duties that the employer is bound to observe (box 25).

Engineering control and housekeeping

Engineering control involves controlling the hazard at the source. The competent authority should ensure that exposure to hazardous substances (such as asbestos, for instance), is prevented or controlled by prescribing engineering controls and work practices which afford maximum protection to workers. One type of engineering control involves built-in protection as part of the work process concerned. These engineering controls should be built in during the design phase. They may be implemented later, but this tends to be more costly. Engineering controls may be more expensive to implement than methods which depend on continual vigilance or intervention by the worker, but they are safer. Examples include machine guarding to prevent accidents or encasing a noise source with a muffler.

Box 25. Duties of employers when workers are exposed to hazards

In situations where workers are exposed to hazards, employers have the duty to:

- inform the workers of all the known hazards associated with their work, the health risks involved and the relevant preventive and protective measures;
- take appropriate measures to eliminate or minimize the risks resulting from exposure to those hazards;
- provide workers with suitable protective equipment, clothing and other facilities where adequate protection against risk of accident or injury to health, including exposure to adverse conditions, cannot be ensured by other means; and
- provide first aid for workers who have suffered from an injury or illness at the workplace, as well as appropriate transportation from the workplace and access to appropriate medical facilities.

Another form of engineering control is the mechanization process. This involves the use of a machine to do dangerous work rather than exposing a worker to the hazard. An example is the use of an automatic parts dipper on a vapour degreaser rather than manually dipping parts into the tank.

Where the elimination of hazardous substances is not practicable in existing plants and processes, employers or managers should apply technical measures to control the hazard or risk by changing the process so as to do the job in a completely different and safer way or by enclosing the process completely to keep the hazard from reaching the worker. If the problems still cannot be solved by these methods, then methods such as local exhaust ventilation could be used to control the hazard.

These and other appropriate measures should be taken so that the exposure level is reduced to a level which, in the light of current knowledge, is not expected to damage the health of workers, even if exposure continues for the duration of working life.

Work practices, including working methods, can ensure that hazardous materials are contained before they become a problem. Where this has failed, strict housekeeping and personal hygiene are absolutely essential to ensure workplace and personal safety. In the presence of toxic chemicals, for instance, strict personal hygiene must always be observed so as to prevent local irritations or the absorption of such chemicals through the skin. Other hazardous substances include lead dust in a storage battery plant or asbestos dust in brake shoe manufacture. Failure to have adequate housekeeping can result in toxic materials circulating in the air. There are several ways of maintaining good housekeeping; for example:

- vacuuming is the best way of cleaning up dust, as dry sweeping often makes the problem worse by pushing dust particles back into the air; and
- regular and thorough maintenance of machines and equipment will reduce dust and fumes.

Substitution

Where necessary for the protection of workers, the competent authority should require the replacement of hazardous substances by substitute materials, in so far as this is possible. For example, in the case of asbestos or products containing asbestos, national laws or regulations must provide for its replacement, if technically practicable, by other materials and products or the use of alternative technology, scientifically evaluated by the competent authority as harmless or less harmful. There could be total or partial prohibition of the use of asbestos or of certain types of asbestos or products containing asbestos in certain work processes. It is, however, necessary to ensure that the substitute is really safer.

Work practices and organizational methods

Where the evaluation of the working environment shows that elimination of risk and total enclosure of machinery are both impracticable, employers should reduce exposure as much as possible, through administrative or organizational measures, so as to:

- reduce the source of the hazard, so that risks are confined to some areas where engineering control measures can be applied effectively;
- adopt adequate work practices and working-time arrangements so that workers' exposure to hazards is effectively controlled; and
- minimize the magnitude of exposure, the number of workers exposed and the duration of exposure, e.g. carry out noisy operations at night or during the weekend, when fewer workers are exposed.

Personal protective equipment

When none of the above approaches is feasible, or when the degree of safety achieved is considered inadequate, the only solution is to provide exposed persons with suitable personal protective equipment and protective clothing. This is the final line of defence and should be used only as a last resort, since it entails reliance on active cooperation and compliance by the workers. Moreover, such equipment may be heavy, cumbersome and uncomfortable, and may restrict movement.

Employers should consult workers or their representatives on suitable personal protective equipment and clothing, having regard to the type of work and risks. Furthermore, when hazards cannot be otherwise prevented or controlled, employers should provide and maintain such equipment and clothing as are reasonably necessary, without cost to the workers. The employer should provide the workers with the appropriate means to enable them to use the individual protective equipment. Indeed, the employer has a duty to ensure its proper use. Protective equipment and clothing should comply with the standards set by the competent authority and take ergonomic principles into account. Workers have the obligation to make proper use of and take good care of the personal protective equipment and protective clothing provided for their use.

Technological change

Technological progress can play an important role in improving working conditions and job content but it can also introduce new hazards. Great care should therefore be taken in both the choice and the international transfer of technology in order to avoid potential hazards and ensure that the technology is adapted to local conditions. There should be consultation between management and workers' representatives whenever new technology is introduced.

The hazards associated with technologies (equipment, substances and processes) used at the work site must be identified and effective measures taken to eliminate or control them. This means that there should be built-in safety factors, and that working conditions, organization and methods should be adapted to the characteristics and capacities of workers.

The introduction of new technology should be accompanied by adequate information and training. Furthermore, potentially dangerous machinery, equipment or substances should not be exported unless adequate safeguards are put in place, including information on safe use in the language of the importing country. It is the duty of the governments of importing countries to review national legislation to make sure that it includes provisions to stop the import of technology detrimental to occupational health and safety or working conditions.

Protection of the general environment

The importance of protecting workers, the general public and the environment from materials containing hazardous substances cannot be overemphasized. In this regard, the competent authority should ensure that criteria consistent with national or international regulations regarding disposal of hazardous waste are established. Procedures to be followed in the disposal and treatment of hazardous waste products should also be established, with a view to ensuring the safety of workers, the protection of the general public and the environment. Employers must therefore dispose of waste containing hazardous materials such as asbestos, in a manner that does not pose a health risk to the workers concerned, including those handling the waste material, or to the general population. Furthermore, it is up to the competent authority and employers to take measures to prevent pollution of the general environment by dust or other pollutants released from the work site.

10. Health promotion, education and training[1]

A healthy, motivated and contented workforce is fundamental to the future social and economic well-being of any nation. To achieve such a workforce, it is not enough to prevent occupational hazards or to protect workers against them. It is also necessary to take positive measures to improve current health status and to promote a health- and safety-oriented culture. Such measures include health promotion, education and training.

Promotion of occupational health and safety

The promotion of occupational health and safety is an organizational investment for the future, because by promoting health in the workplace, enterprises will accrue benefits in the form of reduction in sickness-related costs and an increase in productivity. Consequently, occupational health and safety promotion in the workplace could be regarded as a modern corporate strategy which aims at preventing ill health at work (including work-related diseases, accidents, injuries, occupational diseases and stress) and enhancing the health-promoting potential and well-being of the workforce.

[1] This chapter is based mainly on the Occupational Safety and Health Convention (No. 155), and Recommendation (No. 164), 1981, the Occupational Health Services Recommendation, 1985 (No. 171), the Asbestos Convention, 1986 (No. 162), the Safety and Health in Mines Convention (No. 176), and Recommendation (No. 183), 1995, and PIACT (op. cit.).

As part of national occupational health and safety promotional activities, some countries organize annual awards which are based on certain criteria, including the number of accidents submitted for compensation claims, and continuous hazards inspection and monitoring by the individual workplace. Companies that have kept good safety records are given awards in recognition of their efforts and to encourage others to emulate them. However, mechanisms for ensuring honesty and preventing under-reporting or inaccurate declaration should be put in place and enforced. In other instances, health promotion items, including hazard-monitoring equipment, safety devices, training manuals, and information packages on occupational health and safety are displayed at big annual events such as international trade fairs. Similar activities can be organized at the enterprise level to promote awareness about health and safety. Such activities could include an annual safety festival.

Workers' lifestyles, including diet, exercise, smoking and drinking habits, are a key factor in health. Health education designed to promote good lifestyles and discourage those detrimental to health should be introduced into the workplace as part of the occupational health and safety programme of activities.

Occupational health and safety promotion covers a wide range of measures aimed at increasing interest in working life and occupational health and safety in general. It includes:

- a comprehensive system of information dissemination;
- targeted campaigns for the different sectors of occupational health and safety; and
- safety promotion activities, for example an annual safety week all over the country, featuring events centred around safety themes and culminating in a safety awards ceremony.

The programme on occupational health and safety should include strategies to promote wider awareness of the social and economic importance of improving working conditions and the environment.

An occupational health and safety awareness campaign is aimed at acquainting both management and workers about hazards in their workplaces and their role and obligations in the prevention of occupational accidents, injuries and diseases. It fosters improved communication and work relationships at all levels of the business enterprise including top management, supervisors and workers on the shop floor. It helps a company to achieve its prime objectives, namely, a good health and safety record.

Education in the context of occupational health and safety is designed to communicate a combination of knowledge, understanding and skills that will enable managers and workers in an enterprise to recognize risk factors contributing to occupational accidents, injuries and diseases, and be ready and able to prevent these factors occurring in their own work environment. Occupational health and safety education is thus intended to foster the awareness and positive attitudes which are conducive to health and safety at work.

Education includes training, which is a process of helping others to acquire skills necessary for good performance in a given job. Training is therefore a narrower concept than education. Training, as opposed to full education, may be the only option where workers have limited academic background (hence their comprehension is likely to be limited), or time is scarce.

Education and training provide individuals with the basic theoretical and practical knowledge required for the successful exercise of their chosen occupation or trade. Education and training must therefore also cover the prevention of accidents and injury to health arising out of or linked with or occurring in the course of work. There should be special emphasis on training, including necessary further training. In addition, attention should be paid to the qualifications and motivations of persons involved, in one capacity or another, in the achievement of adequate levels of health and safety.

Where there are health hazards associated with hazardous materials, the competent authority should make appropriate arrangements, in consultation and collaboration with the most representative organizations of employers and workers concerned, to:

- promote the dissemination of information on hazards and on methods of prevention and control; and
- educate all concerned on the hazards and on methods of prevention and control.

Training and information at the national level

The competent authority or authorities in each country should provide information and advice, in an appropriate manner, to employers and workers and promote or facilitate cooperation between them and their organizations, with a view to eliminating hazards or reducing them as far as practicable. Furthermore, where appropriate, a special training programme for migrant workers in their mother tongue should be provided.

Training at all levels should be emphasized as a means of improving working conditions and the work environment. Occupational health and safety institutes and laboratories, labour institutes and other institutions concerned with training, technical support or research in occupational health and safety should be established. Workers' organizations as well as employers should take positive action to carry out training and information programmes with a view to preventing potential occupational hazards in the working environment, and controlling and protecting against existing risks. In their own training, employers should also learn how to gain the confidence of their workers and motivate them; this aspect is as important as the technical content of the training.

The training of labour inspectors, occupational health and safety specialists and others directly concerned with the improvement of working conditions and the work environment should take into account the increasing complexity of work processes. In particular, with the introduction of new or advanced technology, there is a need for training in methods of analysis to identify and measure the hazards, as well as in ways to protect workers against these hazards.

The occupational health and safety programme should place particular emphasis on activities related to the collection, analysis and dissemination of information, taking into consideration the differing needs of government agencies, employers and workers and their organizations, research institutions and others concerned with the improvement of working conditions and the work environment. Priority should be given to the collection and dissemination of information of a practical nature, such as information on provisions of legislation and collective agreements, training activities, research in progress and the content of technical publications.

Information should be easily accessible through a variety of means including the Internet, computerized databases, audiovisual materials, serial publications, information sheets and monographs. A special effort should be made to provide information products at low cost or free of charge to the trade unions and other interested organizations and audiences which might otherwise not be able to afford them.

The establishment of regional, subregional or national information systems on working conditions, occupational health and safety should be encouraged. This could be achieved through the establishment of technical advisory services such as International Occupational Safety and Health Information Centre (CIS) national centres (see box 24), as well as the organization of national and regional workshops and the inclusion of information activities in technical cooperation projects. Information systems should be examined to ensure that there is no overlap with the activities of other institutions providing information in the field of occupational health and safety, and that the most appropriate and cost-effective techniques are used.

Training and information at the enterprise level

The need to give appropriate training in occupational health and safety to workers and their representatives in the enterprise cannot be overemphasized. Training at all levels should be seen as a means of improving working conditions and the work environment. The employers should provide necessary instructions and training, taking account of the functions and capacities of different categories of workers. Furthermore, workers and their representatives should have reasonable time, during paid working hours, to exercise their health and safety functions and to receive training related to them. Employers' and workers' organizations should take positive action to carry out training and information programmes with respect to existing and potential occupational hazards in the work environment. These programmes should focus on:

- prevention;
- control; and
- protection.

Workers should be provided with the type of knowledge commensurate with the technical level of their activity and the nature of their responsibilities. Representatives of workers in the enterprise should also be given adequate information on measures taken by the employer to secure occupational health and safety. They should be able to consult their representative organizations about such information provided that they do not disclose commercial secrets. At an individual level, each worker should be informed in an adequate and appropriate manner of the health hazards involved in his or her work, of the results of the health examinations he or she has undergone and of the assessment of his or her health.

Information activities are a key means of support for occupational health and safety programmes. These activities should emphasize practical materials targeted at specific groups. Special priority should be given to information that can be put to immediate use in enterprises. Policy-makers, labour inspectors and the staff of institutions carrying out research and technical support activities should also be provided with information relevant to their priorities. The participation of such institutions in information networks, both national and international, should be encouraged and developed.

Workers and their health and safety representatives should have access to appropriate information, which might include:

- notice of any forthcoming visits to workplaces by the competent authority in relation to safety or health;
- reports of inspections conducted by the competent authority or the employer, including inspections of machinery or equipment;
- copies of orders or instructions issued by the competent authority in respect of health and safety matters;
- reports of accidents, injuries, instances of ill-health and other occurrences affecting health and safety prepared by the competent authority or the employer;

- information and notices on all hazards at work, including hazardous, toxic or harmful materials, agents or substances used at the workplace;
- any other documentation concerning health and safety that the employer is required to maintain;
- immediate notification of accidents and dangerous occurrences; and
- any health studies conducted in respect of hazards present in the workplace.

Training methods and materials

The importance of training lies in the fact that regulations and warning signs will not prevent risky behaviour unless workers understand dangers and believe that safety measures are worthwhile. Workers, and new recruits in particular, need to be instructed in the safety aspects of their work and kept under close supervision to ensure that they have fully understood the dangers and how to avoid them. This instruction must be supported by effective materials and practical training methods. Specific training materials should be developed to assist action in poorly protected sectors, and emphasis should be placed on the training of trainers.

Developing countries have special needs to which training materials and methods will need to be adapted. In some cases, entirely new materials and methods will be required. This work should utilize research on sectors with particularly high safety risks and pilot experiments identifying the cost-effectiveness and appropriateness of measures. Whenever possible, work on developing training methods and materials should be done in consultation with workers' and employers' representatives.

Given the fact that many workers in developing countries are either illiterate or semi-literate, great care must be taken in choosing an appropriate means of communication. Information on health and safety should be presented in a manner that is easily understood by all workers regardless of their level of education. Language should be kept simple. Everyday language, i.e. the vernacular or local dialect, should be used whenever possible. Information should be conveyed using a medium that does not rely heavily on the written word. Discussions or lectures in the vernacular, along with demonstrations, vivid posters or films are often more effective in putting across health and safety messages. Other techniques include on-the-job demonstrations, role-playing and audiovisual presentations accompanied by explanatory discussions.

Any new techniques implemented must be periodically evaluated. If communication is effective it will produce the desired effects: a reduction in the number of accidents and diseases, or their elimination; savings in medical bills and compensation payments; and improved productivity and worker morale.

Annexes

Annex I. The ILO's SafeWork Programme on Safety, Health and the Environment

Around the world, millions of men and women work in poor and hazardous conditions resulting in some 1.2 million deaths and 160 million workers falling ill each year. It is against this background that the SafeWork Programme on Safety, Health and the Environment has been launched by the ILO. The objectives of SafeWork are to:

- promote preventive policies and programmes;
- extend protection to vulnerable groups of workers;
- better equip governments and employers' and workers' organizations to address workers' well-being, occupational health care and the quality of working life; and
- document the social and economic impact of improving worker protection.

These objectives will be pursued using a four-pronged strategy of advocacy, development of a knowledge base, capacity building for constituents and support for direct action programmes. The main components cover research, policy formulation, training and developing tools needed for programme implementation. Particular attention will be paid to sectors in which risks to life and safety are manifestly high, such as agriculture, mining, transportation and construction as well as workers in the informal sector and those occupationally exposed to abuse and exploitation including women, children and migrants.

SafeWork will be comprehensive, taking into consideration all factors which influence health, safety and productivity. It has been strengthened with the integration of a health promotion component (dealing with substance abuse, stress, violence, etc.) and labour inspection services.

Advocacy, networking and technical cooperation are seen as integral parts of the programme and establish the framework within which a business plan on SafeWork should be built that will truly tap into its potential and maximize the programme's impact on promoting health, well-being, safety and productivity worldwide.

SafeWork aims to create worldwide awareness of the dimensions and consequences of work-related accidents, injuries and diseases; to place the health and safety of all workers on the international agenda; and to stimulate and support practical action at all levels. With this in mind, the programme will launch ground-breaking research, statistical work and media-related activities, and will support national action through a global programme of technical assistance. Human suffering and the cost to society, as well as the potential benefits of protection, such as enhanced productivity, quality and cost savings, will be better documented and publicized. The programme will promote, as a policy and operational tool, the primacy of prevention as an efficient and cost-effective way of providing health and safety protection to all workers.

Strategy

SafeWork will do first things first. It will focus on hazardous work and give primary attention to workers in especially hazardous occupations in sectors where the risks to life and safety are manifestly high, such as agriculture, mining and construction, workers in the informal sector, and those occupationally exposed to abuse and exploitation, such as women, children and migrants.

SafeWork will adopt an integrated approach, including non-traditional aspects of workers' health and safety such as drugs and alcohol, stress and HIV-AIDS. The programme will also make extensive use of gender analysis and planning. There will be strong links within the social protection sector and links with other sectors, InFocus programmes and the field. A key component of SafeWork is its global technical cooperation programme. Partnerships with donors will be strengthened to mobilize additional external resources.

Specific strategies are elaborated below for each of the four goals, and include advocacy, building of the knowledge base, capacity building for constituents and support for direct action programmes.

- **Showing that protection pays.** The prevention of accidents, improvement of working conditions and enforcement of standards are often seen as a cost to business. Little is known about the costs of *not* preventing accidents or poor working conditions, or of the benefits of improvements for productivity and competitiveness. Better information and analytical tools can help increase firms' and governments' willingness to invest in prevention. This strategy will have two main thrusts: extending the knowledge base through a major drive for comprehensive, reliable and sustainable data, and new research on the economics of labour protection. The programme will foster the development of a safety culture worldwide. It will thus demonstrate that prevention policies and programmes benefit all ILO constituents.
- **Protecting workers in hazardous conditions.** Priority must be given to workers in the most hazardous occupations and sectors, such as mining, construction or agriculture, or where working relationships or conditions create particular risks, such as very long working hours, exposure to hazardous

chemicals, work in isolation and work by migrants, etc. The ILO will make use of its extensive experience in the development of standards, codes of practice and technical guides in exploiting the world's information resources, and in developing means of practical action. Member States will be encouraged to set objectives and targets for the protection of workers in hazardous conditions. Particular attention will be given to strengthening the advisory and enforcement capacity of labour inspectorates.

- **Extending protection.** The large majority of workers whose conditions are most in need of improvement are excluded from the scope of existing legislation and other protective measures. Existing policies and programmes need to be reviewed to extend their coverage. This will go hand in hand with action to strengthen labour inspectorates' capacity to develop broad prevention policies and programmes and to promote the protection of vulnerable workers, particularly women workers. Alliances and networks will be extended to include ministries of health, industry, local government, education and social services, as well as local community groups. Emphasis will also be placed on achieving tangible results through practical action and exchanges of information on good practices.
- **Promoting workers' health and well-being.** The strategy to promote workers' health and well-being will involve the establishment of a databank on policies, programmes and good enterprise-level practices so as to improve constituents' capacity to identify workers' protection issues and to provide guidance on new approaches. Governments' capacity for prevention, protection, and the application and enforcement of key labour protection instruments will be strengthened.

SafeWork's projected major outputs

Protecting workers in hazardous jobs

- A *World report on life and death at work*, presenting the world situation regarding risks, accidents and diseases, policies and experience, and guidance for future action
- A film on health and safety, focusing on manifestly hazardous conditions
- New standards on health and safety in agriculture established through tripartite agreement
- A review of standards on occupational health and safety to determine the action needed to update and possibly consolidate them, and to translate them into practical policy and programmatic tools such as codes of practice and guidelines
- Tools and guidance for member States to facilitate the ratification and implementation of ILO standards

- Harmonized chemical labelling systems, safety data sheets and hazard communication methods
- Guidelines for radiation protection and the classification of radiographs of pneumoconiosis
- A rapid response capacity, especially on chemical health and safety issues, including readily accessible networks and timely information

Extending protection to all workers

- Training programmes and tools for owners of small and medium-sized enterprises (SMEs) to promote labour protection and improve productivity
- Strengthening the effectiveness, efficiency and coverage of labour inspection systems
- Guidelines for the extension of labour protection to informal sector workers
- Partnerships with community organizations and others to develop and implement approaches for reaching out to hard-to-reach groups of workers

Promoting workers' health and well-being

- A data bank on policies, programmes and good enterprise-level practices
- Training methodologies and diagnostic tools
- Guidelines on occupational health care for all
- Programmes to prevent and deal with the effects of workplace problems, including drugs, alcohol and stress

Showing that protection pays

- A statistical programme to develop new survey tools and carry out national surveys
- Better national and global estimates of occupational fatalities and injuries
- Report on the economics of accidents and preventive measures
- Tools for inspection services to promote the benefits of prevention
- Guides on occupational health and safety management systems and safety culture
- Tools to reduce work-related environmental damage

Promoting national and industry-based action

- A global technical cooperation programme on safety, health and the environment
- National and industry-level programmes of action to tackle priority issues

Annex II. Glossary

Accumulate: increase, build up.

Acute effect: an immediate, obvious response, usually short-term and often reversible.

Administrative controls: controls designed to limit the amount of time a worker spends at a potentially hazardous job.

Air monitoring: the sampling and measuring of pollutants in the air.

Biological monitoring: usually consists of blood and urine tests performed to look for traces of chemicals and biological indicators of chemical exposure.

Check-list analysis: a method for identifying hazards by comparison with experience in the form of a list of failure modes and hazardous situations.

Code of practice: a document offering practical guidance on the policy, standard-setting and practice in occupational and general public health and safety for use by governments, employers and workers in order to promote health and safety at the national level and the level of the installation. A code of practice is not necessarily a substitute for existing national legislation, regulations and safety standards.

Competent authority: a minister, government department or other public authority with the power to issue regulations, orders or other instructions having the force of law. Under national laws or regulations, the competent authorities may be appointed with responsibilities for specific activities, such as for implementation of national policy and procedures for reporting, recording and notification, workers' compensation, and the elaboration of statistics.

Competent person: a person with suitable training and sufficient knowledge, experience and skill for the performance of the specific work, in good safety conditions. The competent authority may define appropriate criteria for the designation of such persons and may determine the duties to be assigned to them.

Comply: obey (in the case of laws).

Dangerous occurrence: readily identifiable event as defined under national laws and regulations, with potential to cause an injury or disease to persons at work or the public.

Elimination: getting rid of (a specific hazard).

Engineering controls: common control measures, including isolation, enclosure ventilation.

Ergonomic principles: a concept whereby the work to be carried out is organized and specified – and tools and equipment designed and used – in such a way as to be matched with the physical and mental characteristics and capacity of the worker.

Excessive: above the level of comfort.

Exposure: the process of being exposed to something that is around; exposure can affect people in a number of different ways.

Employer: any physical or legal person who employs one or more workers.

Enterprise: an institutional unit or the smallest combination of institutional units that encloses and directly or indirectly controls all necessary functions to carry out its own production activities.

Establishment: an enterprise or part of an enterprise which independently engages in one, or predominantly one, kind of economic activity at or from one location or within one geographic area, for which data are available, or can be meaningfully compiled, that allow the calculation of the operating surplus.

Fatal occupational injury: occupational injury leading to death.

General ventilation: ventilation designed to keep the workplace comfortable.

Hazard: a physical situation with a potential for human injury, damage to property, damage to the environment or some combination of these.

Hazard analysis: the identification of undesired events that lead to the materialization of the hazard, the analysis of the mechanisms by which those undesired events could occur and usually the estimation of the extent, magnitude and relative likelihood of any harmful effects.

Hazard assessment: an evaluation of the results of a hazard analysis including judgements as to their acceptability and, as a guide, comparison with relevant codes, standards, laws and policies.

Hazardous substance: a substance which by virtue of its chemical, physical or toxicological properties constitutes a hazard.

Hazards: dangers.

Housekeeping: keeping the workplace clean and organized.

Hygiene: the practice of principles that maintain health, e.g. cleanliness.

IDLH (Immediately dangerous to life or health): description of an environment that is very hazardous due to a high concentration of toxic chemicals or insufficient oxygen, or both.

Incapacity for work: inability to perform normal duties of work.

Incident: an unsafe occurrence arising out of or in the course of work where no personal injury is caused, or where personal injury requires only first-aid treatment.

Industrial hygiene: the recognition, measurement and control of workplace hazards.

Ingestion: the process of taking a substance into the body through the mouth.

Inhalation: the process of breathing in.

Isolation: an engineering control in which a hazardous job is moved to a place where fewer people will be exposed, or a worker is moved to a place where he or she will not be exposed at all.

Job enrichment: widening of the contents of the work tasks requiring, e.g. higher qualification of the worker.

Job rotation: the worker carries out different work tasks, the change from one task to another occurring according to an agreed system or according to the initiative of the worker's work group.

Job security: security at work against unlawful dismissal, as well as against unsatisfactory work conditions and an unsatisfactory work environment. Sometimes also security against falling income due to sickness or unemployment are included.

Labour inspection: a government function carried out by specially appointed inspectors who regularly visit work sites in order to control whether legislation, rules and regulations are complied with. They normally give verbal and written advice and guidance so as to reduce the risk factors and hazards at the workplace. They should, however, possess and use stronger power, e.g. to stop the work in cases of immediate and serious health and safety hazards or if their advice is repeatedly and unreasonably neglected by the employer. The goal is to improve the work conditions and the work environment.

Labour inspectorate: a government authority with the task of advising and giving directions on issues concerning the protection of workers and the work environment, as well as checking that the protection is sufficient.

Local exhaust ventilation: suction-based ventilation system designed to remove pollutant from the air.

Major accident: an unexpected, sudden occurrence including, in particular, a major emission, fire or explosion, resulting from abnormal developments in the course of an industrial activity, leading to a serious danger to workers, the public or the environment, whether immediate or delayed, inside or outside the installation and involving one or more hazardous substances.

Major hazard installation: an industrial installation which stores, processes or produces hazardous substances in such a form and such a quantity that they possess the potential to cause a major accident. The term is also used for an installation which has on its premises, either permanently or temporarily, a quantity of hazardous substance which exceeds the amount prescribed in national or state major hazard legislation.

Medical surveillance programme: a medical programme, including pre-employment and periodic examinations, which helps to identify early warning signs of occupational diseases.

Monitoring: in the workplace, close observation to determine whether an area is safe for workers.

Non-fatal occupational injury: occupational injury not leading to death.

Notification: procedure specified in national laws and regulations which establishes the ways in which:

- the employer or self-employed person submits information concerning occupational accidents, commuting accidents, dangerous occurrences or incidents; or
- the employer, the self-employed person, the insurance institution or others directly concerned submit information concerning occupational diseases.

Occupational: related to the workplace.

Occupational accident: an occurrence arising out of or in the course of work which results in:

- fatal occupational injury; or
- non-fatal occupational injury.

Occupational disease: a disease contracted as a result of an exposure to risk factors arising from work activity.

Occupational injury: death, any personal injury or disease resulting from an occupational accident.

PIACT: French acronym for the ILO's International Programme for the Improvement of Working Conditions.

Personal hygiene: the practice of principles that maintain personal health, e.g. personal cleanliness.

Personal protective equipment: equipment a worker wears as a barrier between himself or herself and the hazardous agent(s).

Potential hazard: something that may be hazardous.

Recording: procedure specified in national laws and regulations which establish the means by which the employer or self-employed person ensures that information be maintained on:

(a) occupational accidents;

(b) diseases;

(c) commuting accidents; and

(d) dangerous occurrences and incidents.

Reporting: procedure specified by the employer in accordance with national laws and regulations, and in accordance with the practice at the enterprise, for the submission by workers to their immediate supervisor, the competent person, or any other specified person or body, of information on:

(a) any occupational accident or injury to health which arises in the course of or in connection with work;

(b) suspected cases of occupational diseases;

(c) commuting accidents; and

(d) dangerous occurrences and incidents.

Respiratory hazards: hazards to the body's breathing system.

Risk: the likelihood of an undesired event with specified consequences occurring within a specified period or in specified circumstances. It may be expressed either as a frequency (the number of specified events in unit time) or as a probability (the probability of a specified event following a prior event), depending on the circumstances.

Risk management: the whole of actions taken to achieve, maintain or improve the safety of an installation and its operation.

"Safe" levels: levels of exposure to substances below which there will not be a health risk to workers.

Safety audit: a methodical in-depth examination of all or part of a total operating system with relevance to safety.

Safety report: the written presentation of the technical, management and operational information covering the hazards of a major hazard installation and their control in support of a justification for the safety of the installation.

Safety team: a group which may be established by the works management for specific safety purposes, e.g. inspections or emergency planning. The team should include workers, their representatives where appropriate, and other persons with expertise relevant to the tasks.

Self-employed person: as may be defined by the competent authority with reference to the most recent version of the International Classification of Status in Employment (ICSE).

Short-term exposure limit (STEL): the maximum concentration that must not be exceeded for a continuous 15-minute exposure period. STELS are required by law in some countries.

Substitution: replacing particularly hazardous chemicals or work processes by safer ones.

Susceptible: open to hazards, germs, etc.

Time-weighted average (TWA): exposures may be expressed as an 8-hour time-weighted-average (TWA) concentration, which is a measure of exposure intensity that has been averaged over an 8-hour work shift.

Toxic substance: a poisonous substance that can destroy life or injure health.

Vapour: tiny droplets of liquid suspended in the air.

Worker: any person who performs work, either regularly or temporarily, for an employer.

Workers' management: employers and persons at works level having the responsibility and the authority delegated by the employer for taking decisions relevant to the safety of major hazard installations. When appropriate, the definition also includes persons at corporate level having such authority.

Workers' representative: any person who is recognized as such by national law or practice, in accordance with the Workers' Representatives Convention, 1971 (No. 135).

Annex III. Major international labour standards on occupational health and safety

Occupational Safety and Health Convention, 1981 (No. 155)
Excerpts, Articles 1 to 21

PART I. SCOPE AND DEFINITIONS

Article 1

1. This Convention applies to all branches of economic activity.

2. A Member ratifying this Convention may, after consultation at the earliest possible stage with the representative organisations of employers and workers concerned, exclude from its application, in part or in whole, particular branches of economic activity, such as maritime shipping or fishing, in respect of which special problems of a substantial nature arise.

3. Each Member which ratifies this Convention shall list, in the first report on the application of the Convention submitted under article 22 of the Constitution of the International Labour Organisation, any branches which may have been excluded in pursuance of paragraph 2 of this Article, giving the reasons for such exclusion and describing the measures taken to give adequate protection to workers in excluded branches, and shall indicate in subsequent reports any progress towards wider application.

Article 2

1. This Convention applies to all workers in the branches of economic activity covered.

2. A Member ratifying this Convention may, after consultation at the earliest possible stage with the representative organisations of employers and workers

concerned, exclude from its application, in part or in whole, limited categories of workers in respect of which there are particular difficulties.

3. Each Member which ratifies this Convention shall list, in the first report on the application of the Convention submitted under article 22 of the Constitution of the International Labour Organisation, any limited categories of workers which may have been excluded in pursuance of paragraph 2 of this Article, giving the reasons for such exclusion, and shall indicate in subsequent reports any progress towards wider application.

Article 3

For the purpose of this Convention —

(a) the term *branches of economic activity* covers all branches in which workers are employed, including the public service;

(b) the term *workers* covers all employed persons, including public employees;

(c) the term *workplace* covers all places where workers need to be or to go by reason of their work and which are under the direct or indirect control of the employer;

(d) the term *regulations* covers all provisions given force of law by the competent authority or authorities;

(e) the term *health*, in relation to work, indicates not merely the absence of disease or infirmity; it also includes the physical and mental elements affecting health which are directly related to safety and hygiene at work.

PART II. PRINCIPLES OF NATIONAL POLICY

Article 4

1. Each Member shall, in the light of national conditions and practice, and in consultation with the most representative organisations of employers and workers, formulate, implement and periodically review a coherent national policy on occupational safety, occupational health and the working environment.

2. The aim of the policy shall be to prevent accidents and injury to health arising out of, linked with or occurring in the course of work, by minimising, so far as is reasonably practicable, the causes of hazards inherent in the working environment.

Article 5

The policy referred to in Article 4 of this Convention shall take account of the following main spheres of action in so far as they affect occupational safety and health and the working environment:

(a) design, testing, choice, substitution, installation, arrangement, use and maintenance of the material elements of work (workplaces, working environment, tools, machinery and equipment, chemical, physical and biological substances and agents, work processes);

(b) relationships between the material elements of work and the persons who carry out or supervise the work, and adaptation of machinery, equipment, working time, organisation of work and work processes to the physical and mental capacities of the workers;

(c) training, including necessary further training, qualifications and motivations of persons involved, in one capacity or another, in the achievement of adequate levels of safety and health;

(d) communication and co-operation at the levels of the working group and the undertaking and at all other appropriate levels up to and including the national level;

(e) the protection of workers and their representatives from disciplinary measures as a result of actions properly taken by them in conformity with the policy referred to in Article 4 of this Convention.

Article 6

The formulation of the policy referred to in Article 4 of this Convention shall indicate the respective functions and responsibilities in respect of occupational safety and health and the working environment of public authorities, employers, workers and others, taking account both of the complementary character of such responsibilities and of national conditions and practice.

Article 7

The situation regarding occupational safety and health and the working environment shall be reviewed at appropriate intervals, either over-all or in respect of particular areas, with a view to identifying major problems, evolving effective methods for dealing with them and priorities of action, and evaluating results.

PART III. ACTION AT THE NATIONAL LEVEL

Article 8

Each Member shall, by laws or regulations or any other method consistent with national conditions and practice and in consultation with the representative organisations of employers and workers concerned, take such steps as may be necessary to give effect to Article 4 of this Convention.

Article 9

1. The enforcement of laws and regulations concerning occupational safety and health and the working environment shall be secured by an adequate and appropriate system of inspection.

2. The enforcement system shall provide for adequate penalties for violations of the laws and regulations.

Article 10

Measures shall be taken to provide guidance to employers and workers so as to help them to comply with legal obligations.

Article 11

To give effect to the policy referred to in Article 4 of this Convention, the competent authority or authorities shall ensure that the following functions are progressively carried out::

(a) the determination, where the nature and degree of hazards so require, of conditions governing the design, construction and layout of undertakings, the commencement of their operations, major alterations affecting them and changes in their purposes, the safety of technical equipment used at work, as well as the application of procedures defined by the competent authorities;

(b) the determination of work processes and of substances and agents the exposure to which is to be prohibited, limited or made subject to authorisation or control by the competent authority or authorities; health hazards due to the simultaneous exposure to several substances or agents shall be taken into consideration;

(c) the establishment and application of procedures for the notification of occupational accidents and diseases, by employers and, when appropriate, insurance institutions and others directly concerned, and the production of annual statistics on occupational accidents and diseases;

(d) the holding of inquiries, where cases of occupational accidents, occupational diseases or any other injuries to health which arise in the course of or in connection with work appear to reflect situations which are serious;

(e) the publication, annually, of information on measures taken in pursuance of the policy referred to in Article 4 of this Convention and on occupational accidents, occupational diseases and other injuries to health which arise in the course of or in connection with work;

(f) the introduction or extension of systems, taking into account national conditions and possibilities, to examine chemical, physical and biological agents in respect of the risk to the health of workers.

Article 12

Measures shall be taken, in accordance with national law and practice, with a view to ensuring that those who design, manufacture, import, provide or transfer machinery, equipment or substances for occupational use —

(a) satisfy themselves that, so far as is reasonably practicable, the machinery, equipment or substance does not entail dangers for the safety and health of those using it correctly;

(b) make available information concerning the correct installation and use of machinery and equipment and the correct use of substances, and information on hazards of machinery and equipment and dangerous properties of chemical substances and physical and biological agents or products, as well as instructions on how hazards are to be avoided;

(c) undertake studies and research or otherwise keep abreast of the scientific and technical knowledge necessary to comply with subparagraphs (a) and (b) of this Article.

Article 13

A worker who has removed himself from a work situation which he has reasonable justification to believe presents an imminent and serious danger to his life or health shall be protected from undue consequences in accordance with national conditions and practice.

Article 14

Measures shall be taken with a view to promoting in a manner appropriate to national conditions and practice, the inclusion of questions of occupational safety and health and the working environment at all levels of education and training, including higher technical, medical and professional education, in a manner meeting the training needs of all workers.

Article 15

1. With a view to ensuring the coherence of the policy referred to in Article 4 of this Convention and of measures for its application, each Member shall, after consultation at the earliest possible stage with the most representative organisations of employers and workers, and with other bodies as appropriate, make arrangements appropriate to national conditions and practice to ensure the necessary co-ordination between various authorities and bodies called upon to give effect to Parts II and III of this Convention.

2. Whenever circumstances so require and national conditions and practice permit, these arrangements shall include the establishment of a central body.

PART IV. ACTION AT THE LEVEL OF THE UNDERTAKING

Article 16

1. Employers shall be required to ensure that, so far as is reasonably practicable, the workplaces, machinery, equipment and processes under their control are safe and without risk to health.

2. Employers shall be required to ensure that, so far as is reasonably practicable, the chemical, physical and biological substances and agents under their control are without risk to health when the appropriate measures of protection are taken.

3. Employers shall be required to provide, where necessary, adequate protective clothing and protective equipment to prevent, so far is reasonably practicable, risk of accidents or of adverse effects on health.

Article 17

Whenever two or more undertakings engage in activities simultaneously at one workplace, they shall collaborate in applying the requirements of this Convention.

Article 18

Employers shall be required to provide, where necessary, for measures to deal with emergencies and accidents, including adequate first-aid arrangements.

Article 19

There shall be arrangements at the level of the undertaking under which—

(a) workers, in the course of performing their work, co-operate in the fulfilment by their employer of the obligations placed upon him;

(b) representatives of workers in the undertaking co-operate with the employer in the field of occupational safety and health;

(c) representatives of workers in an undertaking are given adequate information on measures taken by the employer to secure occupational safety and health and may consult their representative organisations about such information provided they do not disclose commercial secrets;

(d) workers and their representatives in the undertaking are given appropriate training in occupational safety and health;

(e) workers or their representatives and, as the case may be, their representative organisations in an undertaking, in accordance with national law and practice, are enabled to enquire into, and are consulted by the employer on, all aspects of occupational safety and health associated with their work; for this purpose technical advisers may, by mutual agreement, be brought in from outside the undertaking;

(f) a worker reports forthwith to his immediate supervisor any situation which he has reasonable justification to believe presents an imminent and serious danger to his life or health; until the employer has taken remedial action, if necessary, the employer cannot require workers to return to a work situation where there is continuing imminent and serious danger to life or health.

Article 20

Co-operation between management and workers and/or their representatives within the undertaking shall be an essential element of organisational and other measures taken in pursuance of Articles 16 to 19 of this Convention.

Article 21

Occupational safety and health measures shall not involve any expenditure for the workers.

Occupational Safety and Health Recommendation, 1981 (No. 164)
Excerpts, paragraphs 1 to 17

I. SCOPE AND DEFINITIONS

1. (1) To the greatest extent possible, the provisions of the Occupational Safety and Health Convention, 1981, hereinafter referred to as the Convention, and of this Recommendation should be applied to all branches of economic activity and to all categories of workers.

(2) Provision should be made for such measures as may be necessary and practicable to give self-employed persons protection analogous to that provided for in the Convention and in this Recommendation.

2. For the purpose of this Recommendation —

(a) the term *branches of economic activity* covers all branches in which workers are employed, including the public service;

(b) the term *workers* covers all employed persons, including public employees;

(c) the term *workplace* covers all places where workers need to be or to go by reason of their work and which are under the direct or indirect control of the employer;

(d) the term *regulations* covers all provisions given force of law by the competent authority or authorities;

(e) the term *health*, in relation to work, indicates not merely the absence of disease or infirmity; it also includes the physical and mental elements affecting health which are directly related to safety and hygiene at work.

II. TECHNICAL FIELDS OF ACTION

3. As appropriate for different branches of economic activity and different types of work and taking into account the principle of giving priority to eliminating hazards at their source, measures should be taken in pursuance of the policy referred to in Article 4 of the Convention, in particular in the following fields:

(a) design, siting, structural features, installation, maintenance, repair and alteration of workplaces and means of access thereto and egress therefrom; (b) lighting, ventilation, order and cleanliness of workplaces;

(c) temperature, humidity and movement of air in the workplace;

(d) design, construction, use, maintenance, testing and inspection of machinery and equipment liable to present hazards and, as appropriate, their approval and transfer;

(e) prevention of harmful physical or mental stress due to conditions of work;

(f) handling, stacking and storage of loads and materials, manually or mechanically;

(g) use of electricity;

(h) manufacture, packing, labelling, transport, storage and use of dangerous substances and agents, disposal of their wastes and residues, and, as appropriate, their replacement by other substances or agents which are not dangerous or which are less dangerous;

(i) radiation protection;

(j) prevention and control of, and protection against, occupational hazards due to noise and vibration;

(k) control of the atmosphere and other ambient factors of workplaces;

(l) prevention and control of hazards due to high and low barometric pressures;

(m) prevention of fires and explosions and measures to be taken in case of fire or explosion;

(n) design, manufacture, supply, use, maintenance and testing of personal protective equipment and protective clothing;

(o) sanitary installations, washing facilities, facilities for changing and storing clothes, supply of drinking water, and any other welfare facilities connected with occupational safety and health;

(p) first-aid treatment;

(q) establishment of emergency plans;

(r) supervision of the health of workers.

III. ACTION AT THE NATIONAL LEVEL

4. With a view to giving effect to the policy referred to in Article 4 of the Convention, and taking account of the technical fields of action listed in Paragraph 3 of this Recommendation, the competent authority or authorities in each country should —

(a) issue or approve regulations, codes of practice or other suitable provisions on occupational safety and health and the working environment, account being taken of the links existing between safety and health, on the one hand, and hours of work and rest breaks, on the other;

(b) from time to time review legislative enactments concerning occupational safety and health and the working environment, and provisions issued or approved in pursuance of clause (a) of this Paragraph, in the light of experience and advances in science and technology;

(c) undertake or promote studies and research to identify hazards and find means of overcoming them;

(d) provide information and advice, in an appropriate manner, to employers and workers and promote or facilitate co-operation between them and their organisations, with a view to eliminating hazards or reducing them as far as practicable; where appropriate, a special training programme for migrant workers in their mother tongue should be provided;

(e) provide specific measures to prevent catastrophes, and to co-ordinate and make coherent the actions to be taken at different levels, particularly in industrial zones where undertakings with high potential risks for workers and the surrounding population are situated;

(f) secure good liaison with the International Labour Occupational Safety and Health Hazard Alert System set up within the framework of the International Labour Organisation;

(g) provide appropriate measures for handicapped workers.

5. The system of inspection provided for in paragraph 1 of Article 9 of the Convention should be guided by the provisions of the Labour Inspection Convention, 1947, and the Labour Inspection (Agriculture) Convention, 1969, without prejudice to the obligations thereunder of Members which have ratified these instruments.

6. As appropriate, the competent authority or authorities should, in consultation with the representative organisations of employers and workers concerned, promote measures in the field of conditions of work consistent with the policy referred to in Article 4 of the Convention.

7. The main purposes of the arrangements referred to in Article 15 of the Convention should be to —

(a) implement the requirements of Articles 4 and 7 of the Convention;

(b) co-ordinate the exercise of the functions assigned to the competent authority or authorities in pursuance of Article 11 of the Convention and Paragraph 4 of this Recommendation;

(c) co-ordinate activities in the field of occupational safety and health and the working environment which are exercised nationally, regionally or locally, by public authorities, by employers and their organisations, by workers' organisations and representatives, and by other persons or bodies concerned;

(d) promote exchanges of views, information and experience at the national level, at the level of an industry or that of a branch of economic activity.

8. There should be close co-operation between public authorities and representative employers' and workers' organisations, as well as other bodies concerned in measures for the formulation and application of the policy referred to in Article 4 of the Convention.

9. The review referred to in Article 7 of the Convention should cover in particular the situation of the most vulnerable workers, for example, the handicapped.

IV. ACTION AT THE LEVEL OF THE UNDERTAKING

10. The obligations placed upon employers with a view to achieving the objective set forth in Article 16 of the Convention might include, as appropriate for different branches of economic activity and different types of work, the following:

(a) to provide and maintain workplaces, machinery and equipment, and use work methods, which are as safe and without risk to health as is reasonably practicable;

(b) to give necessary instructions and training, taking account of the functions and capacities of different categories of workers;

(c) to provide adequate supervision of work, of work practices and of application and use of occupational safety and health measures;

(d) to institute organisational arrangements regarding occupational safety and health and the working environment adapted to the size of the undertaking and the nature of its activities;

(e) to provide, without any cost to the worker, adequate personal protective clothing and equipment which are reasonably necessary when hazards cannot be otherwise prevented or controlled;

(f) to ensure that work organisation, particularly with respect to hours of work and rest breaks, does not adversely affect occupational safety and health;

(g) to take all reasonably practicable measures with a view to eliminating excessive physical and mental fatigue;

(h) to undertake studies and research or otherwise keep abreast of the scientific and technical knowledge necessary to comply with the foregoing clauses.

11. Whenever two or more undertakings engage in activities simultaneously at one workplace, they should collaborate in applying the provisions regarding occupational safety and health and the working environment, without prejudice to the responsibility of each undertaking for the health and safety of its employees. In appropriate cases, the competent authority or authorities should prescribe general procedures for this collaboration.

12. (1) The measures taken to facilitate the co-operation referred to in Article 20 of the Convention should include, where appropriate and necessary, the appointment, in accordance with national practice, of workers' safety delegates, of workers' safety and health committees, and/or of joint safety and health committees; in joint safety and health committees workers should have at least equal representation with employers' representatives.

(2) Workers' safety delegates, workers' safety and health committees, and joint safety and health committees or, as appropriate, other workers' representatives should —

(a) be given adequate information on safety and health matters, enabled to examine factors affecting safety and health, and encouraged to propose measures on the subject;

(b) be consulted when major new safety and health measures are envisaged and before they are carried out, and seek to obtain the support of the workers for such measures;

(c) be consulted in planning alterations of work processes, work content or organisation of work, which may have safety or health implications for the workers;

(d) be given protection from dismissal and other measures prejudicial to them while exercising their functions in the field of occupational safety and health as workers' representatives or as members of safety and health committees;

(e) be able to contribute to the decision-making process at the level of the undertaking regarding matters of safety and health;

(f) have access to all parts of the workplace and be able to communicate with the workers on safety and health matters during working hours at the workplace;

(g) be free to contact labour inspectors;

(h) be able to contribute to negotiations in the undertaking on occupational safety and health matters;

(i) have reasonable time during paid working hours to exercise their safety and health functions and to receive training related to these functions;

(j) have recourse to specialists to advise on particular safety and health problems.

13. As necessary in regard to the activities of the undertaking and practicable in regard to size, provision should be made for —

(a) the availability of an occupational health service and a safety service, within the undertaking, jointly with other undertakings, or under arrangements with an outside body;

(b) recourse to specialists to advise on particular occupational safety or health problems or supervise the application of measures to meet them.

14. Employers should, where the nature of the operations in their undertakings warrants it, be required to set out in writing their policy and arrangements in the field of occupational safety and health, and the various responsibilities exercised under these arrangements, and to bring this information to the notice of every worker, in a language or medium the worker readily understands.

15. (1) Employers should be required to verify the implementation of applicable standards on occupational safety and health regularly, for instance by environmental monitoring, and to undertake systematic safety audits from time to time.

(2) Employers should be required to keep such records relevant to occupational safety and health and the working environment as are considered necessary by the competent authority or authorities; these might include records of all notifiable occupational accidents and injuries to health which arise in the course of or in connection with work, records of authorisation and exemptions under laws or regulations to supervision of the health of workers in the undertaking, and data concerning exposure to specified substances and agents.

16. The arrangements provided for in Article 19 of the Convention should aim at ensuring that workers —

(a) take reasonable care for their own safety and that of other persons who may be affected by their acts or omissions at work;

(b) comply with instructions given for their own safety and health and those of others and with safety and health procedures;

(c) use safety devices and protective equipment correctly and do not render them inoperative;

(d) report forthwith to their immediate supervisor any situation which they have reason to believe could present a hazard and which they cannot themselves correct;

(e) report any accident or injury to health which arises in the course of or in connection with work.

17. No measures prejudicial to a worker should be taken by reference to the fact that, in good faith, he complained of what he considered to be a breach of statutory requirements or a serious inadequacy in the measures taken by the employer in respect of occupational safety and health and the working environment.

Occupational Health Services Convention, 1985 (No. 161)
Excerpts, Articles 1 to 15

PART I. PRINCIPLES OF NATIONAL POLICY

Article 1

For the purpose of this Convention —

(a) the term *occupational health services* means services entrusted with essentially preventive functions and responsible for advising the employer, the workers and their representatives in the undertaking on —

 (i) the requirements for establishing and maintaining a safe and healthy working environment which will facilitate optimal physical and mental health in relation to work;

 (ii) the adaptation of work to the capabilities of workers in the light of their state of physical and mental health;

(b) the term *workers' representatives in the undertaking* means persons who are recognised as such under national law or practice.

Article 2

In the light of national conditions and practice and in consultation with the most representative organisations of employers and workers, where they exist, each Member shall formulate, implement and periodically review a coherent national policy on occupational health services.

Article 3

1. Each Member undertakes to develop progressively occupational health services for all workers, including those in the public sector and the members of production co-operatives, in all branches of economic activity and all undertakings. The provision made should be adequate and appropriate to the specific risks of the undertakings.

2. If occupational health services cannot be immediately established for all undertakings, each Member concerned shall draw up plans for the establishment of such services in consultation with the most representative organisations of employers and workers, where they exist.

3. Each Member concerned shall indicate, in the first report on the application of the Convention submitted under article 22 of the Constitution of the International Labour Organisation, the plans drawn up pursuant to paragraph 2 of this Article, and indicate in subsequent reports any progress in their application.

Article 4

The competent authority shall consult the most representative organisations of employers and workers, where they exist, on the measures to be taken to give effect to the provisions of this Convention.

PART II. FUNCTIONS

Article 5

Without prejudice to the responsibility of each employer for the health and safety of the workers in his employment, and with due regard to the necessity for the workers to participate in matters of occupational health and safety, occupational health services shall have such of the following functions as are adequate and appropriate to the occupational risks of the undertaking:

(a) identification and assessment of the risks from health hazards in the workplace;

(b) surveillance of the factors in the working environment and working practices which may affect workers' health, including sanitary installations, canteens and housing where these facilities are provided by the employer;

(c) advice on planning and organisation of work, including the design of workplaces, on the choice, maintenance and condition of machinery and other equipment and on substances used in work;

(d) participation in the development of programmes for the improvement of working practices as well as testing and evaluation of health aspects of new equipment;

(e) advice on occupational health, safety and hygiene and on ergonomics and individual and collective protective equipment;

(f) surveillance of workers' health in relation to work;

(g) promoting the adaptation of work to the worker;

(h) contribution to measures of vocational rehabilitation;

(i) collaboration in providing information, training and education in the fields of occupational health and hygiene and ergonomics;

(j) organising of first aid and emergency treatment;

(k) participation in analysis of occupational accidents and occupational diseases.

PART III. ORGANISATION

Article 6

Provision shall be made for the establishment of occupational health services —

(a) by laws or regulations; or

(b) by collective agreements or as otherwise agreed upon by the employers and workers concerned; or

(c) in any other manner approved by the competent authority after consultation with the representative organisations of employers and workers concerned.

Article 7

1. Occupational health services may be organised as a service for a single undertaking or as a service common to a number of undertakings, as appropriate.

2. In accordance with national conditions and practice, occupational health services may be organised by —

(a) the undertakings or groups of undertakings concerned;

(b) public authorities or official services;

(c) social security institutions;

(d) any other bodies authorised by the competent authority;

(e) a combination of any of the above.

Article 8

The employer, the workers and their representatives, where they exist, shall cooperate and participate in the implementation of the organisational and other measures relating to occupational health services on an equitable basis.

PART IV. CONDITIONS OF OPERATION

Article 9

1. In accordance with national law and practice, occupational health services should be multidisciplinary. The composition of the personnel shall be determined by the nature of the duties to be performed.

2. Occupational health services shall carry out their functions in co-operation with the other services in the undertaking.

3. Measures shall be taken, in accordance with national law and practice, to ensure adequate co-operation and co-ordination between occupational health services and, as appropriate, other bodies concerned with the provision of health services.

Article 10

The personnel providing occupational health services shall enjoy full professional independence from employers, workers, and their representatives, where they exist, in relation to the functions listed in Article 5.

Article 11

The competent authority shall determine the qualifications required for the personnel providing occupational health services, according to the nature of the duties to be performed and in accordance with national law and practice.

Article 12

The surveillance of workers' health in relation to work shall involve no loss of earnings for them, shall be free of charge and shall take place as far as possible during working hours.

Article 13

All workers shall be informed of health hazards involved in their work.

Article 14

Occupational health services shall be informed by the employer and workers of any known factors and any suspected factors in the working environment which may affect the workers' health.

Article 15

Occupational health services shall be informed of occurrences of ill health amongst workers and absence from work for health reasons, in order to be able to identify whether there is any relation between the reasons for ill health or absence and any health hazards which may be present at the workplace. Personnel providing occupational health services shall not be required by the employer to verify the reasons for absence from work.

Occupational Health Services Recommendation, 1985 (No. 171)
Excerpts, paragraphs 1 to 43

I. PRINCIPLES OF NATIONAL POLICY

1. Each Member should, in the light of national conditions and practice and in consultation with the most representative organisations of employers and workers, where they exist, formulate, implement and periodically review a coherent national policy on occupational health services, which should include general principles governing their functions, organisation and operation.

2. (1) Each Member should develop progressively occupational health services for all workers, including those in the public sector and the members of production co-operatives, in all branches of economic activity and all undertakings. The provision made should be adequate and appropriate to the specific health risks of the undertakings.

(2) Provision should also be made for such measures as may be necessary and reasonably practicable to make available to self-employed persons protection analogous to that provided for in the Occupational Health Services Convention, 1985, and in this Recommendation.

II. FUNCTIONS

3. The role of occupational health services should be essentially preventive.

4. Occupational health services should establish a programme of activity adapted to the undertaking or undertakings they serve, taking into account in particular the occupational hazards in the working environment as well as the problems specific to the branches of economic activity concerned.

A. SURVEILLANCE OF THE WORKING ENVIRONMENT

5. (1) The surveillance of the working environment should include —

(a) identification and evaluation of the environmental factors which may affect the workers' health;

(b) assessment of conditions of occupational hygiene and factors in the organisation of work which may give rise to risks for the health of workers;

(c) assessment of collective and personal protective equipment;

(d) assessment where appropriate of exposure of workers to hazardous agents by valid and generally accepted monitoring methods;

(e) assessment of control systems designed to eliminate or reduce exposure.

(2) Such surveillance should be carried out in liaison with the other technical services of the undertaking and in co-operation with the workers concerned and their representatives in the undertaking or the safety and health committee, where they exist.

6. (1) In accordance with national law and practice, data resulting from the surveillance of the working environment should be recorded in an appropriate manner and be available to the employer, the workers and their representatives in the undertaking concerned or the safety and health committee, where they exist.

(2) These data should be used on a confidential basis and solely to provide guidance and advice on measures to improve the working environment and the health and safety of workers.

(3) The competent authority should have access to these data. They may only be communicated by the occupational health service to others with the agreement of the employer and the workers or their representatives in the undertaking or the safety and health committee, where they exist.

7. The surveillance of the working environment should entail such visits by the personnel providing occupational health services as may be necessary to examine the factors in the working environment which may affect the workers' health, the environmental health conditions at the workplace and the working conditions.

8. Occupational health services should —

(a) carry out monitoring of workers' exposure to special health hazards, when necessary;

(b) supervise sanitary installations and other facilities for the workers, such as drinking water, canteens and living accommodation, when provided by the employer;

(c) advise on the possible impact on the workers' health of the use of technologies;

(d) participate in and advise on the selection of the equipment necessary for the personal protection of the workers against occupational hazards;

(e) collaborate in job analysis and in the study of organisation and methods of work with a view to securing a better adaptation of work to the workers;

(f) participate in the analysis of occupational accidents and occupational diseases and in accident prevention programmes.

9. Personnel providing occupational health services should, after informing the employer, workers and their representatives, where appropriate —

(a) have free access to all workplaces and to the installations the undertaking provides for the workers;

(b) have access to information concerning the processes, performance standards, products, materials and substances used or whose use is envisaged, subject to their preserving the confidentiality of any secret information they may learn which does not affect the health of workers;

(c) be able to take for the purpose of analysis samples of products, materials and substances used or handled.

10. Occupational health services should be consulted concerning proposed modifications in the work processes or in the conditions of work liable to have an effect on the health or safety of workers.

B. SURVEILLANCE OF THE WORKERS' HEALTH

11. (1) Surveillance of the workers' health should include, in the cases and under the conditions specified by the competent authority, all assessments necessary to protect the health of the workers, which may include —

(a) health assessment of workers before their assignment to specific tasks which may involve a danger to their health or that of others;

(b) health assessment at periodic intervals during employment which involves exposure to a particular hazard to health;

(c) health assessment on resumption of work after a prolonged absence for health reasons for the purpose of determining its possible occupational causes, of recommending appropriate action to protect the workers and of determining the worker's suitability for the job and needs for reassignment and rehabilitation;

(d) health assessment on and after the termination of assignments involving hazards which might cause or contribute to future health impairment.

(2) Provisions should be adopted to protect the privacy of the workers and to ensure that health surveillance is not used for discriminatory purposes or in any other manner prejudicial to their interests.

12. (1) In the case of exposure of workers to specific occupational hazards, in addition to the health assessments provided for in Paragraph 11 of this Recommendation, the surveillance of the workers' health should include, where appropriate, any examinations and investigations which may be necessary to detect exposure levels and early biological effects and responses.

(2) When a valid and generally accepted method of biological monitoring of the workers' health for the early detection of the effects on health of exposure to specific occupational hazards exists, it may be used to identify workers who need a detailed medical examination, subject to the individual worker's consent.

13. Occupational health services should be informed of occurrences of ill health amongst workers and absences from work for health reasons, in order to be able to identify whether there is any relation between the reasons for ill health or absence and any health hazards which may be present at the workplace. Personnel providing occupational health services should not be required by the employer to verify the reasons for absence from work.

14. (1) Occupational health services should record data on workers' health in personal confidential health files. These files should also contain information on jobs held by the workers, on exposure to occupational hazards involved in their work, and on the results of any assessments of workers' exposure to these hazards.

(2) The personnel providing occupational health services should have access to personal health files only to the extent that the information contained in the files is relevant to the performance of their duties. Where the files contain personal information covered by medical confidentiality this access should be restricted to medical personnel.

(3) Personal data relating to health assessments may be communicated to others only with the informed consent of the worker concerned.

15. The conditions under which, and time during which, personal health files should be kept, the conditions under which they may be communicated or transferred

and the measures necessary to keep them confidential, in particular when the information they contain is placed on computer, should be prescribed by national laws or regulations or by the competent authority or, in accordance with national practice, governed by recognised ethical guide-lines.

16. (1) On completing a prescribed medical examination for the purpose of determining fitness for work involving exposure to a particular hazard, the physician who has carried out the examination should communicate his conclusions in writing to both the worker and the employer.

(2) These conclusions should contain no information of a medical nature; they might, as appropriate, indicate fitness for the proposed assignment or specify the kinds of jobs and the conditions of work which are medically contra-indicated, either temporarily or permanently.

17. Where the continued employment of a worker in a particular job is contra-indicated for health reasons, the occupational health service should collaborate in efforts to find alternative employment for him in the undertaking, or another appropriate solution.

18. Where an occupational disease has been detected through the surveillance of the worker's health, it should be notified to the competent authority in accordance with national law and practice. The employer, workers and workers' representatives should be informed that this notification has been carried out.

C. INFORMATION, EDUCATION, TRAINING, ADVICE

19. Occupational health services should participate in designing and implementing programmes of information, education and training on health and hygiene in relation to work for the personnel of the undertaking.

20. Occupational health services should participate in the training and regular retraining of first-aid personnel and in the progressive and continuing training of all workers in the undertaking who contribute to occupational safety and health.

21. With a view to promoting the adaptation of work to the workers and improving the working conditions and environment, occupational health services should act as advisers on occupational health and hygiene and ergonomics to the employer, the workers and their representatives in the undertaking and the safety and health committee, where they exist, and should collaborate with bodies already operating as advisers in this field.

22. (1) Each worker should be informed in an adequate and appropriate manner of the health hazards involved in his work, of the results of the health examinations he has undergone and of the assessment of his health.

(2) Each worker should have the right to have corrected any data which are erroneous or which might lead to error.

(3) In addition, occupational health services should provide workers with personal advice concerning their health in relation to their work.

D. FIRST AID, TREATMENT AND HEALTH PROGRAMMES

23. Taking into account national law and practice, occupational health services in undertakings should provide first-aid and emergency treatment in cases of accident or indisposition of workers at the workplace and should collaborate in the organisation of first aid.

24. Taking into account the organisation of preventive medicine at the national level, occupational health services might, where possible and appropriate —

(a) carry out immunisations in respect of biological hazards in the working environment;

(b) take part in campaigns for the protection of health;

(c) collaborate with the health authorities within the framework of public health programmes.

25. Taking into account national law and practice and after consultation with the most representative organisations of employers and workers, where they exist, the competent authority should, where necessary, authorise occupational health services, in agreement with all concerned, including the worker and his own doctor or a primary health care service, where applicable, to undertake or to participate in one or more of the following functions:

(a) treatment of workers who have not stopped work or who have resumed work after an absence;

(b) treatment of the victims of occupational accidents;

(c) treatment of occupational diseases and of health impairment aggravated by work;

(d) medical aspects of vocational re-education and rehabilitation.

26. Taking into account national law and practice concerning the organisation of health care, and distance from clinics, occupational health services might engage in other health activities, including curative medical care for workers and their families, as authorized by the competent authority in consultation with the most representative organisations of employers and workers, where they exist.

27. Occupational health services should co-operate with the other services concerned in the establishment of emergency plans for action in the case of major accidents.

E. OTHER FUNCTIONS

28. Occupational health services should analyse the results of the surveillance of the workers' health and of the working environment, as well as the results of biological monitoring and of personal monitoring of workers' exposure to occupational hazards, where they exist, with a view to assessing possible connections between exposure to occupational hazards and health impairment and to proposing measures for improving the working conditions and environment.

29. Occupational health services should draw up plans and reports at appropriate intervals concerning their activities and health conditions in the undertaking. These plans and reports should be made available to the employer and the workers' representatives in the undertaking or the safety and health committee, where they exist, and be available to the competent authority.

30. (1) Occupational health services, in consultation with the employers' and the workers' representatives, should contribute to research, within the limits of their resources, by participating in studies or inquiries in the undertaking or in the relevant branch of economic activity, for example, with a view to collecting data for epidemiological purposes and orienting their activities.

(2) The results of the measurements carried out in the working environment and of the assessments of the workers' health may be used for research purposes, subject to the provisions of Paragraphs 6(3), 11(2) and 14(3) of this Recommendation.

31. Occupational health services should participate with other services in the undertaking, as appropriate, in measures to prevent its activities from having an adverse effect on the general environment.

III. ORGANISATION

32. Occupational health services should, as far as possible, be located within or near the place of employment, or should be organised in such a way as to ensure that their functions are carried out at the place of employment.

33. (1) The employer, the workers and their representatives, where they exist, should co-operate and participate in the implementation of the organisational and other measures relating to occupational health services on an equitable basis.

(2) In conformity with national conditions and practice, employers and workers or their representatives in the undertaking or the safety and health committee, where they exist, should participate in decisions affecting the organisation and operation of these services, including those relating to the employment of personnel and the planning of the service's programmes.

34. (1) Occupational health services may be organised as a service within a single undertaking or as a service common to a number of undertakings, as appropriate.

(2) In accordance with national conditions and practice, occupational health services may be organised by —

(a) the undertakings or groups of undertakings concerned;

(b) the public authorities or official services;

(c) social security institutions;

(d) any other bodies authorised by the competent authority;

(e) a combination of any of the above.

(3) The competent authority should determine the circumstances in which, in the absence of an occupational health service, appropriate existing services may, as an interim measure, be recognised as authorised bodies in accordance with subparagraph 2(d) of this Paragraph.

35. In situations where the competent authority, after consulting the representative organisations of employers and workers concerned, where they exist, has determined that the establishment of an occupational health service, or access to such a service, is impracticable, undertakings should, as an interim measure, make arrangements, after consulting the workers' representatives in the undertaking or the safety and health committee, where they exist, with a local medical service for carrying out the health examinations prescribed by national laws or regulations, providing surveillance of the environmental health conditions in the undertaking and ensuring that first-aid and emergency treatment are properly organised.

IV CONDITIONS OF OPERATION

36. (1) In accordance with national law and practice, occupational health services should be made up of multidisciplinary teams whose composition should be determined by the nature of the duties to be performed.

(2) Occupational health services should have sufficient technical personnel with specialised training and experience in such fields as occupational medicine, occupational hygiene, ergonomics, occupational health nursing and other relevant fields. They should, as far as possible, keep themselves up to date with progress in the scientific and technical knowledge necessary to perform their duties and should be given the opportunity to do so without loss of earnings.

(3) The occupational health services should, in addition, have the necessary administrative personnel for their operation.

37. (1) The professional independence of the personnel providing occupational health services should be safeguarded. In accordance with national law and practice, this might be done through laws or regulations and appropriate consultations between the employer, the workers, and their representatives and the safety and health committees, where they exist.

(2) The competent authority should, where appropriate and in accordance with national law and practice, specify the conditions for the engagement and termination of employment of the personnel of occupational health services in consultation with the representative organisations of employers and workers concerned.

38. Each person who works in an occupational health service should be required to observe professional secrecy as regards both medical and technical information which may come to his knowledge in connection with his functions and the activities of the service, subject to such exceptions as may be provided for by national laws or regulations.

39. (1) The competent authority may prescribe standards for the premises and equipment necessary for occupational health services to exercise their functions.

(2) Occupational health services should have access to appropriate facilities for carrying out the analyses and tests necessary for surveillance of the workers' health and of the working environment.

40. (1) Within the framework of a multidisciplinary approach, occupational health services should collaborate with —

(a) those services which are concerned with the safety of workers in the undertaking;

(b) the various production units, or departments, in order to help them in formulating and implementing relevant preventive programmes;

(c) the personnel department and other departments concerned;

(d) the workers' representatives in the undertaking, workers' safety representatives and the safety and health committee, where they exist.

(2) Occupational health services and occupational safety services might be organised together, where appropriate.

41. Occupational health services should also, where necessary, have contacts with external services and bodies dealing with questions of health, hygiene, safety, vocational rehabilitation, retraining and reassignment, working conditions and the welfare of workers, as well as with inspection services and with the national body which has been designated to take part in the International Occupational Safety and Health Hazard Alert System set up within the framework of the International Labour Organisation.

42. The person in charge of an occupational health service should be able, in accordance with the provisions of Paragraph 38, to consult the competent authority, after informing the employer and the workers' representatives in the undertaking or the safety and health committee, where they exist, on the implementation of occupational safety and health standards in the undertaking.

43. The occupational health services of a national or multinational enterprise with more than one establishment should provide the highest standard of services, without discrimination, to the workers in all its establishments, regardless of the place or country in which they are situated.

Annex IV. Example of a health and safety policy[1]

Brent and Harrow Health Authority (United Kingdom)

HEALTH AND SAFETY POLICY

1. Introduction

Brent and Harrow Health Authority, as an employer, recognizes and accepts its responsibility, under the Management of Health and Safety at Work Regulations 1992, for providing a safe and healthy workplace and working environment for all its employees. It shall be the policy of the Authority to promote the highest standards of health, safety and welfare to ensure the prevention of injuries. This will be achieved by the promotion of occupational health and hygiene, the control of all situations likely to cause injury to persons and property, fire prevention and fire control and by the provision of appropriate safety training.

2. General policy

The Authority will implement the requirements of the Management of Health and Safety at Work Regulations 1992 and will, therefore:

- in conjunction with staff representatives, assess the risk to the health and safety of its employees and to anyone else who may be affected by its activity, so that necessary preventive and protective measures can be identified;
- implement the preventive and protective measures that follow from the risk assessment;
- provide appropriate health surveillance of employees where necessary;
- set up emergency procedures in conjunction with staff representatives;
- give employees information about health and safety matters;

[1] The example given is not an endorsement of the policy on the part of the ILO and should not be considered as a perfect model. Furthermore, it is not considered to be representative of all undertakings, as it may not be applicable to small and medium-sized enterprises.

- cooperate with any other employers who share a work site;
- provide information to subcontractors to the Authority;
- ensure that employees have adequate health and safety training and are capable enough at their jobs to avoid risk.

3. Implementation

3.1 Risk assessment

The authority shall carry out risk assessment for its sites. The assessment will involve the identification of hazards present (whether arising from work activities or from other factors, e.g. the layout of the premises) and then evaluating the extent of the risk. These assessments will be carried out on an annual basis in conjunction with the staff representatives.

In addition, risk assessments can be called by staff or management side representatives for issues which require urgent consideration.

The risk assessments shall:

- ensure that all significant risks or hazards are addressed;
- address what actually happens in the workplace or during work activity;
- ensure that all groups of employees and others who might be affected are considered;
- identify groups of workers who might particularly be at risk;
- take account of existing preventative or precautionary measures;
- be carried out on a regular basis and the significant findings recorded.

The record of the significant findings shall include:

- the significant hazards identified in the assessment;
- the existing control measures in place and the extent to which they control the risks;
- the population which may be affected by these significant risks or hazards, including any groups who are particularly at risk.

3.2 Health and safety management

The Authority will put into place arrangements to manage health and safety issues. This will involve a systematic approach which identifies priorities and sets objectives. Wherever possible, risks will be eliminated by the careful selection and design of facilities and equipment. The Authority will strive for progressive improvements in health and safety and will, through the assessment process, monitor and review progress against set objectives.

3.3 **Health surveillance**

The Authority shall ensure that all its employees are provided with health surveillance, as is appropriate, taking into consideration the risks identified through the assessment process. This will include provision for eye-testing in accordance with the Health and Safety (Display Screen Equipment) Regulations 1992 and the provison of pre-employment screening provided by the Occupational Health Service.

3.4 **Health and safety officer**

The Authority shall identify individual officers to assist with health and safety measures. The Authority will ensure that each individual is competent to carry out whatever tasks he/she is assigned and given adequate information and support.

3.5 **Staff consultation**

The Authority will consult with staff representatives to ensure that the objectives of this policy are met. This will include the development of a subcommittee of the Authority's Staff Consultative body to participate in risk assessments, monitor implementation plans and to assist in the testing of emergency procedures.

3.6 **Information for employees**

The risk assessments will help to identify information which has to be provided to employees. Relevant information on risks and on preventive and protective measures will be limited to what employees and others need to know to ensure their health and safety. In addition, information will be provided on the emergency procedures outlined in the previous section, including the identity of staff nominated to assist in the event of an evacuation.

3.7 **Training**

The risk assessment will help to identify the level of training needed for each type of work as part of the preventive and protective measures. This can include basic skills training, specific on-the-job training and training in health and safety or emergency procedures. New employees shall receive basic induction training on health and safety, including arrangements for first aid, fire and evacuation.

3.8 Employees' duties

Employees have a duty to take reasonable care for their own health and safety and that of others who may be affected by their acts. Employees should use correctly all work items provided by the Authority, in accordance with the training they receive to enable them to use the items safety. Employees are also required to cooperate with the Authority to enable it to comply with its statutory duties. The Authority should be made aware without delay of any work situation which might present a serious and imminent danger. Employees should also notify the Authority of any shortcomings in the health and safety arrangements even when no immediate danger exists, so that remedial action can be taken as may be needed.

3.9 Procedures for serious and imminent danger

The Authority will establish procedures to be followed by any worker if situations presenting serious and imminent danger were to arise. The procedures will provide clear guidance on when employees and others should stop work and move to a place of safety. The risk assessment should identify foreseeable events which need to be covered by these procedures.

These procedures will be tested on an annual basis in conjunction with staff representatives.

3.10 Cooperation and coordination

Where the Authority shares premises with one or more other employers, it will cooperate to ensure that all obligations are met. This will include coordination of measures which cover the workplace as a whole to be fully effective. The designated health and safety officer(s) will take lead responsibility for these activities.

Annex V. Bibliography

ILO codes of practice

Accident prevention on board ship at sea and in port, 2nd edn. (1996)

Ambient factors at the workplace (2001)

HIV/AIDS and the world of work (forthcoming)

Management of alcohol and drug-related issues in the workplace (1996)

Occupational exposure to airborne substances harmful to health (1991)

Occupational safety and health in the iron and steel industry (1983)

Prevention of accidents due to electricity underground in coal mines (1959)

Prevention of accidents due to explosions underground in coal mines (1974)

Prevention of accidents due to fires underground in coal mines (1959)

Prevention of major industrial accidents (1991)

Protection of workers' personal data (1997)

Radiation protection of workers (ionising radiations) (1987)

Recording and notification of occupational accidents and diseases (1996)

Safe construction and installation of electric passenger, goods and service lifts (1972)

Safe construction and operation of tractors (1976)

Safe design and use of chain saws (1978)

Safety and health in building and civil engineering work (1985)

Safety and health in coal mines (1986)

Safety and health in construction (1992)

Safety and health in dock work, 2nd edn. (1992)

Safety and health in forestry work, 2nd edn. (1998)

Safety and health in opencast mines (1991)

Safety and health in shipbuilding and ship repairing (1984) [out of print, but may be consulted in libraries]

Safety and health in the construction of fixed offshore installations in the petroleum industry (1982)

Safety, health and working conditions in the transfer of technology to developing countries (1988)

Safety in the use of asbestos (1990)

Safety in the use of chemicals at work (1993)

Safety in the use of synthetic, vitreous fibre insulation wools (glass wool, rock wool, slag wool) (2001)

Other ILO publications

Bakar Che Man, A. and D. Gold: *Safety and health in the use of chemicals at work: A training manual* (Geneva, 1993).

Bureau for Workers' Activities (ACTRAV). *Your health and safety at work: A modular training package* (Geneva, 1996).

Di Martino, V. and N. Corlett (eds.): *Work organization and ergonomics* (Geneva, 1998).

Encyclopaedia of occupational health and safety, 4th edition, edited by Jeanne Mager Stellman, 4 Vols. (Geneva, 1998).

Ergonomic checkpoints: Practical and easy-to-implement solutions for improving safety, health and working conditions, revised edition (Geneva, 1999).

Guidelines on occupational safety and health management systems (forthcoming).

Major hazard control: A practical manual (Geneva, 1994).

N'Daba, L. and J. Hodges-Aeberhard (eds.): *HIV/AIDS and employment* (Geneva, 1998).

Protection of workers from power frequency electric and magnetic fields: A practical guide, Occupational Safety and Health Series, No. 69 (Geneva, 1993).

Safety, health and welfare on construction sites: A training manual (Geneva, 1995).

Safety in the use of radiofrequency dielectric heaters and sealers: A practical guide, Occupational Safety and Health Series, No. 71 (Geneva, 1998).

Technical and ethical guidelines for workers' health surveillance, Occupational Safety and Health Series, No. 72 (Geneva, 1999).

The organization of first aid in the workplace, Occupational Safety and Health Series, No. 63 (Geneva, 1989).

The use of lasers in the workplace: A practical guide, Occupational Safety and Health Series, No. 68 (Geneva, 1993).

Visual display units: Radiation protection guidance, Occupational Safety and Health Series, No. 70 (Geneva, 1993).

Index

Note: Italic page numbers refer to text boxes; bold numbers refer to annexes.